POLLUTION: CAUSES, EFFECTS AND CONTROL

Special Publication No. 44

Pollution: Causes, Effects and Control

Based on the Papers Given at a Residential School, Organised by the Continuing Education Committee of The Royal Society of Chemistry

University of Lancaster, 13th-15th September, 1982

Edited by
Roy M. Harrison
University of Lancaster

The Royal Society of Chemistry
Burlington House, London W1V 0BN

British Library Cataloguing in Publication Data

Pollution.—(Special publication Royal Society of Chemistry, ISSN 0260-6291; no. 44)
1. Pollution—Congresses
I. Harrison, Roy M. II. Series
363.7'3 TD174

ISBN 0-85186-875-4

Printed in Great Britain by
Whitstable Litho Ltd., Whitstable, Kent

Introduction

Pollution is an inevitable consequence of Man's existence on earth. As such, there will always be a requirement for the scientific study of the pollutants themselves, their biological effects, and for engineering controls within a suitable legislative framework. Recent years have seen intense activity within all of these areas which has accompanied an increased public awareness of pollution problems.

Pollution is a truly inter-disciplinary subject area in which few practitioners have received formal training in more than one of the many disciplines contributing to the knowledge and understanding of pollution problems. For this reason there is much to be gained from a broader perspective of the subject.

The chapters in this book are in the main derived from the course notes provided by lecturers at an R.S.C. Residential School on this topic held at Lancaster University in September, 1982. These have been supplemented by a few additional contributions aimed at improving the overall coverage of this very broad subject area. The Residential School had a teaching function and the chapters are pitched at a level appropriate to this objective, rather than reviewing the most up-to-date research results. Nonetheless, the knowledge and insights provided by the lecturers do, in my view, provide a valuable overview of the field which will be of benefit to all those involved in the pollution field.

Roy M. Harrison

Contents

1

The Control of Industrial Pollution

By A. K. Barbour
RIO TINTO-ZINC CORPORATION PLC, YORK HOUSE, BOND STREET, BRISTOL BS1 3PE, U.K.

1. What is Pollution?

This may seem a very naive question with which to start a presentation to a Royal Society of Chemistry Summer School, but the varying answers which are given by persons from different backgrounds and perspectives can and do provoke much confusion and misunderstanding. Academic debates around these differences between "industrialists" and "environmentalists" or "conservationists" - (Why can we not appreciate that the two are often the same?) - are often reflected in regulatory circles, particularly the EEC.

Some will argue that any chemical, biological or biochemical change from some natural "baseline", often regrettably imaginary, constitutes "pollution". In other words, "pollution" is thought of as any departure from what is perceived to be the natural order of things. By this definition, "pollution" is often thought of as an inescapable concomitant of "development" or "civilisation". "Nature" does not pollute, "industry" does - "natural" products are better than "synthetics", even if they are chemically identical, natural lifestyles are more satisfying and healthier than modern "civilisation".

Others will argue that fears that industrial development is proceeding at such a pace that it is "out of control" are totally unfounded. The professionals within "government" and "industry", prodded from time to time by indications of "public opinion", will control all problems if only they are left to themselves to get on with it. I incline to this view myself.

As is usual with controversial subjects of this kind, the "truth" (so far as it can be perceived) lies somewhere in between the above-stated extreme positions. It is, in my view, unfortunate (but inescapable) that in these environmental debates, the known scientific facts at any time become overlain with emotive opinion, speculation, political and socio-economic aspects. But we have to face the fact that the process industries, and the scientists and other professionals who work in them, operate in the real world where the non-scientific aspects are important and often appreciated and understood much more thoroughly than the scientific bases. They are also much easier to talk about for most people!

This talk will thus concentrate on the scientific aspects of industrial pollution control, partly because this is what I am best qualified to present, but mainly because, whilst in the end these aspects may not be the determining factors, I believe they are the most important. If the scientific database, or its interpretation, is wrong we are likely to take the wrong overall decisions - and, in this respect, environmental questions are little different from any others having a significant scientific content.

Returning therefore to the opening question, my definition of "pollution" is "any degree of contamination of air, water or land which is likely to produce a significant adverse health effect to a significant number of persons in a foreseeable period of time". You will see that my definition contains no absolutes and several compromises. In my view, emissions do not necessarily constitute "pollution". There is a degree of contamination of the environment which can be accepted because it provokes no significant adverse health effect. In other words, natural absorptive and degradation processes come into play. Such effects have to be felt by a significant number of persons; it is impracticable to protect completely every single human being, including all highly-sensitive persons. There is ample scope here for debate both on the above propositions themselves and on quantifying the word "significant". There is also much room for closer definition of the grey areas between a real adverse health effect and a nuisance. Questions of this kind frequently become important in relation to visual and aesthetic issues (e.g. the actual appearance of a dump for mining waste on previously flat land or in a valley), noise emissions and malodour questions.

2. Chemicals and Concentrations

How often do we hear phrases such as "Chemicals are poisoning our Environment" or "Factories spewing out poisons" - they never emit, they usually spew! Or, in the slightly more rarified atmosphere of some scientific meetings - or even reports from some respected academic institutions, "Cadmium (or Lead or Mercury) is highly toxic"........so ban Cadmium pigments, or Lead, or Mercury-based chlorine electrolysis cells. Concentrations of the chemical in question are rarely stated; in the case of heavy metals the chemical compound or species is almost never stated.

Even the best-designed laboratory experiments on plants or animals, using regular application of soluble chemicals, are difficult to relate to effects in real life where exposures are variable, concentrations usually very much lower and chemical species almost certainly quite different. We must always bear in mind these difficulties, and avoid developing hypothesis - which is always necessary to explain experimental results and to plan future work - to the point where it becomes speculation having no experimental basis.

But if we are going to control pollution effectively - and in this context that means industrial pollution - environmental priorities have to be determined. To do this we must establish the following with reasonable accuracy:-

- concentration and mass emission levels over significant time periods.
- likely health effects, including food chain effects, of the <u>chemical species</u> under review. This will involve both <u>detailed toxicology</u> and <u>exposure estimations</u>.
- essential and desirable levels of concentration and mass for control for both "out-plant" and "in-plant" exposures.
- identification of <u>effective production systems</u> which will manufacture the product at the determined "out-plant" and "in-plant" criteria.

  Note that it is desirable, but usually only practicable with new plant, to consider the <u>production process as a whole</u>. Environmental protection should be considered as an integral part of the total process, not something to be achieved by the separate addition of arrestment equipment, effluent treatment plants, etc.

3. Toxicology and Control

My judgement is that it is clearly correct to separate biological effects from control strategies There is clearly a distinct and fundamentally vital activity dealing with the toxicology, in all its aspects, of the identified chemical species which are considered to be candidate pollutants. Once this has been done, it is necessary to make an accurate estimation of the effects likely to result from the exposure, at various levels, of human beings to the toxic chemical.

Turning now to the strategy for control, it is in my view essential for emission criteria to be set at levels which can be met routinely by well designed and managed practical operating plants. For this reason, I believe strongly in the UK system for Air Pollution control wherein the Inspectorate determines emission levels taking into account expert opinion and advice both on health and exposure effects, and levels which can practicably be achieved in operating plants. Other UK regulatory systems for Environmental Control follow similar precepts to a somewhat lesser degree. On the "in-plant" side the UK Factory Inspectorate currently follows the criteria published by the American Committee of Governmental Industrial Hygienists (ACGIH) which, in turn, considers both health and control aspects in setting its Threshold Limit Values* (TLV), Short-term Exposure Limits* (STEL) and Ceiling Values.

Essential to the successful working of a system of this type is the existence of an experienced corps of Inspectors who understand industrial conditions and who can accurately prejudge the performance of ventilation systems, arrestment equipment, etc., in the field.

---

*TLV-TWA : "the time-weighted average concentration for a normal 8-hour workday or 40-hour workweek, to which nearly all workers may be repeatedly exposed, day after day, without adverse effect."

*TLV-STEL : "the maximal concentration to which workers can be exposed for a period up to 15 minutes continuously without suffering from (i) irritation (ii) chronic or irreversible tissue change, or (iii) narcosis of sufficient degree to increase accident proneness, impair self-rescue or materially reduce work efficiency, provided that no more than four excursions per day are permitted with at least 60 minutes between exposure periods and that the daily TLV-TWA is not exceeded."

*TLV-C : "the concentration that should not be exceeded even instantaneously."

4. Toxicological Evaluation

The Health and Safety Commission recently estimated that some 20 to 30,000 separate chemicals are manufactured world-wide on a scale of greater than 1 tonne per annum and that no fewer than 4 million new compounds all told were synthesised during the last decade. Thus it is not altogether surprising that more rigorous and subtle toxicological testing is revealing the disconcerting fact that some well-known chemicals of commerce are much more hazardous, or have different toxicological effects, than was previously known. In the following list asbestos is now known to be a highly potent lung carcinogen, lead may affect the mental development of children, cadmium can accumulate in liver and kidney and cause disfunction of both organs, vinyl chloride may cause liver cancer - all effects which have been determined during the last decade or so.

Some Recent Problems in Toxicological Assessment

- Asbestos
- Lead
- Cadmium
- Arsenic
- Benzene
- Vinyl chloride
- Formaldehyde
- Acrylonitrile

Additionally, many other commonplace chemicals are coming under review as suspect carcinogens by regulatory agencies in both the USA and the EEC.

International Agency for Research on Cancer (IARC) (1979)

Unequivocally carcinogenic for humans

17 chemicals + 2 processes including:

- Arsenic and arsenic compounds
- 4-Amino biphenyl
- Benzene
- bis (Chloromethyl) ether
- Mustard gas
- Soots, tars + oils
- Asbestos
- Benzidine
- Chromium + some chromium compounds
- β-naphthylamine
- Vinyl chloride

Probably carcinogenic to humans

6 chemicals with higher degree of evidence including:

- Aflatoxins
- Cadmium + certain cadmium compounds
- Nickel + certain nickel compounds

12 chemicals with lower degree of evidence including:

- Acrylonitrile
- Beryllium + certain beryllium compounds
- Carbon tetrachloride
- Dimethyl sulphate
- Ethylene oxide
- P.C.B.'s
- Phenacetin

5. Standards Setting

In an area such as toxicology, where the scientific basis is developing rapidly, and which relates directly to human health, there is an inevitable tendency for assessments to be rushed and for some of the discussions to become emotive, even confrontational. Differing standards may develop in different countries, e.g. for lead, benzene, arsenic and asbestos, I believe because the regulatory authorities have varying health objectives.

Occupational Lead Standards in Various Countries

	Pb_B (μg/100ml)	Pb_A (μg/ m^3)
UK	80 (men) 40 (women)	150 (8 hr.av.)
EEC	60-70 (men) 45 (women)	100 (40 hr.av.)
USA	50 (after May '83) 40 (return level)	50 (max)
Japan	60	150 (max)

Some seem to take the extreme position that the concentration of pollutants inside a plant should be no different from environmental concentrations - a laudable objective but currently quite impossible of achievement for many pollutants.

Where clearly-defined adverse health effects are directly related to workplace exposure, it is correct to set standards of exposure which will prevent these effects, whether or not technology exists to meet them. In effect, this would mean that many plants responsible for these effects would be closed. But where such clearly-defined adverse health effects are absent, it is in my view unnecessary, even irresponsible, for regulatory agencies to promulgate standards which are unnecessarily stringent for the protection of worker health, particularly where they are unattainable by any known technology.

Responsible people in the regulatory agencies, management and the unions can and should reach practicable and uniform standards for worker protection based on a consensus of underlying philosophy. The experience of the relevant Industrial Medical

Officers must be utilised fully in establishing these standards - perhaps this body of practical experience is under-utilised at the present time.

As noted earlier, it is particularly difficult to establish standards for suspect carcinogens where there is a bewildering range of test methods and interpretations of results. Current controversy embraces the relevance of animal testing (dose, mode of entry, period of test) and the significance and methodology of epidemiological evaluation. Of the 2000 or so chemicals shown to have some degree of carcinogenic potential in animal tests, only a few tens are known to be human carcinogens. Further evidence of current confusion in this field is that EEC studies suggest that occupational cancers account for 1-3% of all cancers, a figure in the same general range as that derived by Chemical Industry Association specialists. This figure, however, contrasts sharply with the 20-40% recently publicised by ASTMS in their recent booklet on the subject.

In the welter of argument and controversy which currently surrounds these questions it is easy to overlook the tremendous reductions which have been made in the exposure of workpeople to toxicants at work in the last 10-20 years. Where the protection of worker health is concerned there is a natural tendency to "shoot low", i.e. to set standards which are more rigorous than is justified by perceived health effects. This must be resisted - though it must be admitted that establishment of "correct" standards in this area is a task of great difficulty and complexity.

My personal "order of priority" can be summarised as follows:

<u>Environmental Standards - The Order of Priority</u>

- What should the standard be?
  - Toxicology
  - Biology
  - Scale of production/distribution
  - Degree of exposure
- How to achieve it routinely?
  - Technology
  - Engineering
  - Work practices

Always be prepared to change and adapt to new information

6. The Regulatory Framework for Controlling Pollution

(i) Air Pollution

In the UK the Alkali Inspectorate system for air pollution control is based on the following:

Best Practicable Means

Practicable — "Reasonably practicable having regard, amongst other things, to local conditions and circumstances, to the financial implications and to the current state of technical knowledge."

Means — "Design, installation, maintenance, manner and periods of operations of plant and machinery, and the design, construction and maintenance of buildings."

In general, emission standards are not specified in UK legislation; "Best Practicable Means" for each individual emission are ultimately expressed by the Alkali Inspectorate as so-called "Presumptive Limits" which are based upon the following criteria:

Alkali Inspectorate Criteria for Presumptive Limits

1. No emission can be tolerated which constitutes a demonstrable health hazard, either short or long term.
2. Emissions, in terms of both concentration and mass, must be reduced to the lowest practicable amount.
3. Having secured the minimum practicable emission, the height of discharge must be arranged so that the residual emission is rendered harmless and inoffensive. (For highly toxic metals, the concentration of each source to the existing background concentration shall not exceed one-fortieth of the Threshold Limit Value for a factory atmosphere on a three-minute mean basis. In deciding on the most important parameter, the effects on vegetation, animals and amenity are also considered.)

It will be seen, therefore, that implementation of the above criteria provides a fully satisfactory level of health protection for persons living in the proximity of major point sources of pollution, consistent with the best current medical knowledge at the time. Two points need to be emphasised. One is that TLV's or Threshold Limit Values (that concentration which can be tolerated by an average person exposed eight hours per day for forty hours per week for prolonged periods) are currently set by an independent group of professional specialists (the American Conference of Government Industrial Hygienists) and adopted without modification by the UK Factory Inspectorate. The ACGIH is, of course, totally independent of either the Alkali Inspectorate or the Factory Inspectorate. The second point concerns definition of the phrase "demonstrable health hazard, either short or long term". It must be emphasised that this is a moving target, reflected usually by lower (or more stringent) TLV's adopted by the ACGIH in the light of increasing toxicological and biochemical knowledge. Scientists can usually accept as normal that increasing knowledge can result in changing standards; the general public often finds great difficulty in grasping the idea that current knowledge is far from complete at any given time and hence that standards are not fixed for all time.

The detailed future of the UK system for controlling air pollution, like many other environmental issues, may be influenced to a greater or lesser degree by developments in the EEC. In particular, there is evidence from the Smoke/Sulphur dioxide directive, and that on lead, that EEC thinking will follow that in the USA in establishing general ambient air guidelines first and then attempting to derive fixed stack emission values from the ambient air guidelines.

Many of the practical difficulties which have beset US attempts to control air pollution have been attributed to rigorous adherence to this system. Nevertheless, the 5th Report of the Royal Commission recognised the value of Ambient Air Guidelines - not least of which was the importance of the general public's "right to know" the quality of the air which they breathe. There is little doubt in my mind that some element of control by Ambient Standards will be built into future Alkali Inspectorate regulations - as noted earlier, this has already happened for SO_2/smoke and lead - but it is very much to be hoped that control of individual point sources will remain the foundation of our system.

(ii) Liquid Effluent Control

For liquid effluents the philosophy underlying the Environmental Quality Objective (EQO) approach favoured by the UK is to categorise the receiving waters into end-uses (e.g. potable, recreational, industrial, estuarial, etc.) and then to set pollutant concentrations in the receiving waters consistent with the natural survival of "critical organisms" appropriate for that end-use. Individual effluent standards are then derived so as to be consistent with these receiving water criteria.

Control of Aqueous Effluents by Environmental Quality (EQO/EQS)

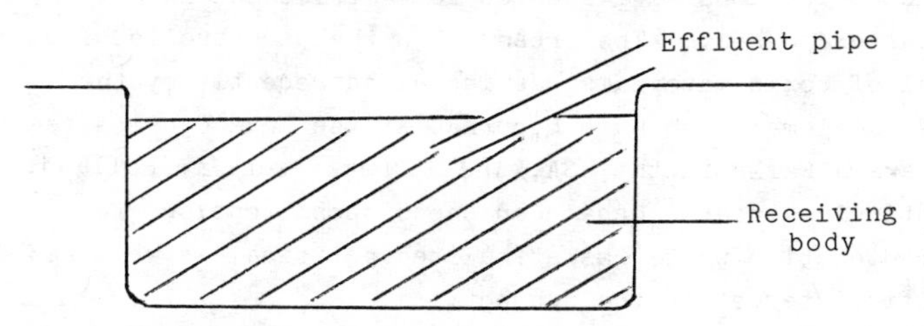

1. Use of receiving body of water:
   - Potable
   - Recreational
   - Industrial/Agricultural
   - Estuarial, etc.
2. Define critical organism and the tolerable concentration of pollutant for its survival.
3. Calculate corresponding concentration of pollutant in the effluent pipe.
4. Monitor both the pipe and the receiving body.
5. Alter if justified by monitoring.

This approach contrasts with some overseas legislation (e.g. EEC) which seeks to impose national fixed standards for the concentration of pollutants in effluents regardless of the end-use category of the receiving waters. In the extreme, effluents discharging into estuaries would have to be purified to the same level as those discharging into fresh water streams used for potable waters. For the so-called "black list" pollutants (including mercury and cadmium) specified in the Dangerous Substances in the Aquatic Environment Directive, the UK is currently setting up quantitative EQO-based schemes designed to be at least as effective environmentally as the EEC fixed standards approach without incurring the expense and misallocation of resources which it implies.

(iii) Disposal of Solid Wastes

Somewhat surprisingly, regulatory control over the disposal of potentially hazardous solid wastes dates essentially from the early 1970's when the UK passed the Deposit of Poisonous Wastes Act, 1972, controlling disposal to land, and the Paris and Oslo Conventions were adopted to restrict the deposit of solid wastes to sea. The pressure on the uncontrolled land disposal of toxic materials has increased recently by the serious problems which have occurred at the Love Canal site near Niagara Falls in the USA, and at Lekkerkerk in Holland. The continuing intense debate on the disposal options for "hot" radioactive waste has stimulated political as well as scientific interest in the subject.

Control of solid waste disposal in the UK is vested mainly with the County Councils (and the GLC) operating under policy guidelines laid down by the Department of the Environment centrally. These Waste Disposal Authorities are required to discuss with the Regional Water Authorities the question of leachate from landfills so that contamination of ground water and other water systems is avoided. The private Waste Disposal Contractors provide an important service in co-operation with the WDA's: the majority of hazardous wastes are disposed through contractors operating under the Site Licensing Control exercised through the WDA's.

The Deposit of Poisonous Wastes Act, 1972 (the provisions of which are now to be superseded by Section 17 of the Control of Pollution Act, 1974) made it an offence to create an environmental hazard by the deposition on land of any solid other than those which are specifically excluded. It also established a mechanism by which the intention to deposit such a waste is notified in advance to both the local Waste Disposal Authority and the Water Authority. The new legislation represented by Section 17 of the Control of Pollution Act provides a much smaller inclusive list of defined hazardous wastes which required special control.

By and large, the Deposit of Poisonous Wastes Act has worked reasonably well in practice and my personal view is that the new regulations will also be found to be effective and operable without undue difficulty. However, there are some potential problems in the area of land disposal of wastes:

(i) In general, UK philosophy is 'dilute and disperse' and recent work particularly at Harwell has provided this concept with a firm scientific foundation in many cases. However, both US and German authorities seem to prefer to concentrate toxic wastes in sealed sites which will undoubtedly require long-term supervision and control.

(ii) In an administrative sense, it is likely that an increasing number of local authorities will show reluctance to accept industrial solid wastes under some pretext or another.

The use of high-security sites (i.e. clay lined sites, deep injection, and/or encapsulation or chemical fixation) should, as now, be restricted to the disposal of highly hazardous materials or to situations where groundwater contamination is probable.

In addition we should continue to be able to use the option of sea-disposal wherever it is sensible scientifically to do so. The sea has enormous absorptive powers and these must continue to be available for the disposal of toxic wastes in a controlled and rational manner. The Oslo Convention of 1972, to which the UK was an early signatory, is fortunately administered in a balanced manner by MAFF in the UK but nevertheless - in my view - represents a piece of international regulation on a subject which was (and is) a problem only in a few localised areas.

8. Process Needs for Environmental Protection

There is no doubt that many modern processes, developed in the post World War II period through to, say, 1970, originally for reasons based on labour savings, raw materials availability or economy in capital cost, are intrinsically capable of meeting higher environmental standards than the processes they replaced. Some examples in the chemical and metallurgical process industries are:

Better Process Technology Aids Environmental Protection

Copper	replacement of reverberatory furnaces by continuous processes permits efficient sulphur dioxide capture as sulphuric acid
Zinc	much reduced in- and out-plant metal concentrations by replacing distillation processes by electrolytic and blast furnace processes
Aluminium	fluoride emissions greatly reduced by dry or wet scrubbing of pot-line gases
Sulphuric Acid	"DCDA" (Double Contact, Double Absorption) technology reduces sulphur dioxide stack concentration to 500 ppm
Vinyl Chloride	in-plant concentrations greatly reduced; acetylene raw material eliminated
Chlorine	modern diaphragm process could eliminate mercury in effluents

Paradoxically, in the decade which has seen a marked heightening of environmental awareness, process Research and Development in industry has declined substantially. It has to be admitted that this is largely because new process development has failed, for a variety of reasons, to show clearly the return required in these recessionary times. Nevertheless, from the environmental as well as the economic standpoint, this decline is much to be regretted - particularly in cases where, for "resource" reasons, environmentally "difficult" processes are considered justified. Examples of the latter are the use of coal instead of oil for chemicals production, some examples of metals recycling, or the generation of power from coal.

The EEC has now recognised this problem in its proposed "Cleaner Technology" programme, and it is greatly to be hoped that satisfactory arrangements can be made for industrial scientists and engineers to participate fully in this programme so that any technology emerging from it is suitable for industrial application.

9. Training

In an area fraught with so much uncertainty, frequently compounded by emotive publicity, and of such potential cost to industry, it is clearly vital that the right people in industry should be considering these problems. They must be properly trained, in detail and depth, to identify problems and develop practical solutions.

In my view, we need environmental awareness to be built into the existing technical disciplines, and in some disciplines, particularly toxicology, occupational hygiene, and industrial medicine, we need more fully qualified practitioners. In other words, we need a synthesis or bringing together of existing disciplines, co-ordinated by a few experienced generalists. In industry at least (and it may be different in Central and Local Government) we do not need the large numbers of people for whom places now exist in the numerous recently created "environmental science" or "environmental studies" courses. We need masters of singularly difficult, individual disciplines, not jacks of all trades.

Thus, on the process side, implementation of environmental improvement involves:

- the design, maintenance and operation of plant and associated ventilation systems which will create and sustain a safe working plant atmosphere (i.e. one in which known pollutants are present, at concentrations below the current Threshold Limit Value);
- the establishment of adequate systems for the sampling and analysis of in-plant atmospheres and interpretation of the results;
- with an increasing number of pollutants, the medical and biochemical monitoring of the health of process operators, to ensure that operating and ventilation regimes are effective;
- monitoring of the out-plant environment to ensure that pollutant concentrations meet Alkali Inspectorate Criteria (or equivalent) both at the stack and in the area of maximum impact on the ground.

Already we see that several specialist disciplines are required if industry is to meet these obligations:

- Industrial Hygienists to set the pollution control standards for the in-plant environment, to establish a control strategy and to monitor the results and suggest improvements where necessary;

- Mechanical and Electrical Engineers to design both the plant itself and its associated ventilation, arrestment and effluent systems;

- Industrial Medical Officers to monitor the health of the work force and, in conjunction with industrial hygienists, to identify areas for improvement on the plant;

- Chemists to set up and operate both the in-plant and out-plant monitoring schemes and to develop new analytical and sampling methods as necessary.

As well as the above actions necessary to preserve a safe in-plant working environment, most processes have to be designed so that:

- Emissions to Atmosphere meet Alkali Inspectorate Criteria;

- Liquid Effluents meet the purity criteria laid down by the Regional Water Authority;

- Solid Wastes, either from the process directly, or material accumulated as a result of purifying gaseous or liquid streams, must be disposed of in an environmentally acceptable manner.

Each of these areas has developed in recent years to be almost a discipline in its own right - certainly each has developed a cadre of authoritative specialists. The control of emissions to atmosphere clearly requires a thorough knowledge of arrestment systems such as scrubbing towers, venturis, cyclones, driers, bag-plants, electro-static precipitators and the like. But it also requires a detailed knowledge of the variation in process conditions through the complete process cycle and also the start-up, shut-down and emergency procedures. A critical aspect of this part of plant environmental design is often the balance between process air and ventilation air and whether it is advantageous to eject through a single, usually very tall stack, or several shorter stacks. Whilst the former is usually the ideal, plant lay-outs often militate against the single stack concept requiring long runs of costly and difficult-to-maintain trunk mains around the plants.

Liquid effluent treatment is, again, a specialised discipline. It requires a good knowledge of those basic chemical principles which relate to the purification of solutions by techniques such as:

- removal of oxygen-demanding substances;
- precipitation followed by thickening and/or filtration;
- ion-exchange;
- solvent extraction;
- reverse osmosis;

backed up by the chemical engineer's appreciation of the equipment involved and the electrical engineer's knowledge of the controls which are required to maintain the necessary high standard of day and night operation.

The disposal of solid wastes may be a direct consequence of the process itself (e.g. the irony red-muds which arise during the production of alumina from bauxite or titanium dioxide from ilmenite, slags from blast furnaces, waste from metalliferous ores concentration), or from the purification of liquid effluents and the dedusting of gaseous streams. In the former cases, the arisings may represent quite considerable tonnages and separate tips or, for tailings, specially-designed lagoons may be justified. In any case, dump stability will have to be assured, some knowledge of dump chemistry will have to be assembled and the probable fate of streams assessed from the hydro-geological point of view to ensure that undesirable contamination of water resources does not occur. Of course, a person dealing with solid wastes will have to be conversant with all methods of handling large tonnages of such materials and to establish a good knowledge of transportation costs.

All three of these aspects of waste disposal have ecological implications whether it be the effect of dust or gaseous fall-out on the local farming community, the effect of pollutants in a liquid effluent on "target" organisms in rivers, or the effect of tip leachate on local water resources. For these "out-plant" aspects of process environmental control, we can thus add specialist chemists and chemical engineers, specialists in mechanical handling, biologists in both agricultural and marine systems, and geologists.

Training along traditional lines for such disciplines seems to me to provide the best basis for our environmental manpower needs.

Society's interest in and concern about environmental matters poses, and will continue to pose, substantial questions for the scientific, industrial and legislative communities. Many of these questions will require continuing expenditures on research and development to provide adequate scientific and technological answers. In my view, industrialised societies will not accept the reduction in living standards which would inevitably follow from the major reductions in industrial activity implied by a "zero growth" policy or a return to agrarianism. Industry, Society and legislators must work together to equilibrate reasonable industrial development with environmental standards which are scientifically defensible, practicably operable and acceptable to Society at large.

2
Water Quality and Health

By R. F. Packham
WATER RESEARCH CENTRE, MEDMENHAM LABORATORY, PO BOX 16, HENLEY ROAD, MEDMENHAM, MARLOW, BUCKS. SL7 2HD, U.K.

INTRODUCTION

The main consideration in ensuring the safety of public water supplies is the removal of bacteriological contamination. Major epidemics of water borne disease in the UK were only eliminated when

(a) their bacteriological origin was recognised,
(b) the sewage contamination of water supplies was minimised,
(c) disinfection treatment was introduced.

There have been no major epidemics since the Croydon typhoid epidemic of 1937 following which all water supplies were chlorinated.

Concern about possible health effects of chemicals in drinking water stemmed from

1. Recognition that environmental factors are involved in many diseases. Drinking water is a part of the environment.
2. Application of sophisticated analytical techniques which revealed the presence in drinking water of traces of many chemicals. Some of these would cause concern at high concentrations.
3. Public interest in some specific pollution problems e.g. detergents, pesticides, lead, asbestos.

Acute health effects of chemical constituents of drinking water are very unlikely but there could be effects due to exposure to low concentrations over a lifetime. Such evidence as there is suggests that any such effects are extremely small but this is still the subject of much research.

DRINKING WATER STANDARDS

In the UK, public water supplies must be 'wholesome'. There is no official definition of wholesome but WHO drinking water standards[1,2] have been widely referred to. An EC Directive on the quality of water for human consumption[3] has however now been promulgated and this will have to be implemented in member states in 1985. The Directive sets 'maximum acceptable concentrations' (MAC) for about 40 determinands but gives no rationale for these limits. Also of interest are revised WHO standards or 'Guidelines for Drinking Water Quality' as they are now called. These will be published as two volumes including rationale by early 1983 and will give 'guideline values' for nine inorganics and eighteen organics together with rationale.

All standards rightly give prime emphasis to the avoidance of bacteriological contamination. Many of the limits for chemicals are precautionary concentrations that are rarely approached in practice. There are only a few for which there is any substantial evidence that health effects have occurred in practice due to consumption of water.

CURRENT ISSUES IN THE UK

1. Situations where there is difficulty in complying with established limits for certain water constituents e.g. lead and nitrate.
2. Situations where there is statistical evidence associating disease incidence with drinking water quality but where the causal element has not been identified e.g. inverse relationship between cardiovascular mortality and water hardness.
3. Situations where there is concern about possible health effects due to water contaminants without any real evidence either way e.g. possible health effects due to waste water recycling (re-use) or due to byproducts of disinfection including trihalomethanes.

LEAD

Lead in drinking water is almost entirely due to lead household plumbing, usually but not invariably when in combination with soft water. WHO (European) Standards give an upper limit for lead in drinking water of

100 μg/l or 300 μg/l after 16 hours contact with lead pipes. The EC Drinking Water Directive gives a MAC of 50 μg/l but an annotation indicates that this applies to a sample taken after flushing. It adds that if the sample is taken either directly or after flushing and the lead content exceeds 100 μg/l either frequently or to an appreciable extent "suitable measures must be taken to reduce the exposure to lead on the part of the consumer".

Although it has been shown that exposure to low concentrations of lead in drinking water can increase blood lead levels there is considerable controversy over the significance of this to health. Concern centres particularly on possible neurophysiological effects influencing learning ability and general behaviour in children. The report of the Lawther Committee[4] concluded that there is no convincing evidence of deleterious effects of blood lead concentration below about 35 μg/dl but there is pressure to revise this in the light of more recent evidence.

A survey in Great Britain of lead in drinking water in 1975[5] gave the results shown in Table 1.

Table 1. Lead in drinking water (DOE 1975 survey)

	Percentage of households with lead concentrates		
Sample	>50 μg/l	> 100 μg/l	>300 μg/l
Morning first draw	20.4	9	1.6
Random daytime	10.3	4.3	0.9

Following the survey, water authorities have undertaken more detailed surveys to locate areas where there are problems in meeting standards.

Remedial measures include

1. Changing the water supply to one that is non-plumbosolvent - rarely possible.
2. Water treatment - pH adjustment and sometimes orthophosphate addition. This is the most cost effective remedy when practical.
3. Remove lead plumbing and instal more suitable material. Totally effective but very expensive - about £600 per house - on average.

4. Tell householder to flush to tap before taking water for drinking purposes - useful interim measure.

NITRATES

Nitrate levels in groundwater have increased since the 1960s due to many factors - changed land use, increased use of nitrogenous fertilizers and in rivers, increased recycling of sewage effluent. Limit for nitrate in drinking water is based on its effect on a blood disease in bottle fed infants - methaemaglobinaemia. There is also concern about possible effects on the incidence of stomach cancer. Methaemaglobinaemia is rare in UK even in high nitrate areas and epidemiological evidence suggests little effect, if any, on gastric cancer in UK[6]. Evidence is accumulating that water nitrate can only affect these diseases when there is malnutrition - particularly vitamin C deficiency.

WHO (European) desirable limit is 50 mg/l (as NO_3) but up to 100 mg/l is acceptable if medical authorities are warned about the possible danger of infantile methaemaglobinaemia. EC Directive gives a MAC of 45 mg/l but there may be grounds for derogation in UK.

Although this level is exceeded particularly in groundwater in several parts of England there is no evidence that it represents a health problem. In some cases levels can be reduced by blending. Treatment techniques based on ion exchange or biological denitrification have been developed but so far these have only been used on a pilot scale in this country.

WATER QUALITY AND CARDIOVASCULAR DISEASE

The hardness of water supplies in Britain tends to follow a North to South East gradient with the softest supplies in Scotland, Northern England and Wales and the hardest in East Anglia and Southern England. Mortality from cardiovascular disease (heart disease and stroke) tends to follow the same pattern and several statistical studies have demonstrated an inverse relationship between cardiovascular disease and water hardness.

The following conclusions were drawn from the Regional Heart Study[7] the largest and most recent of these studies.

(i) For the 253 towns in England, Wales and Scotland having a population greater than 50,000 at the 1971 Census, there is a highly significant inverse relationship between the hardness of drinking water and mortality from cardiovascular disease. This relationship persists when age, sex, socioeconomic and climatic effects are taken into account, and is not shown for mortality from non-cardiovascular diseases.

(ii) On average, very soft-water towns have about 10% higher cardiovascular mortality than medium-hard or harder water towns, after adjustment for socioeconomic and climatic factors. There is, however, considerable scatter in the data, as can be seen from Figure 1. The relationship also appears to be non-linear in that most of the variation in mortality appears to take place at the soft end of the hardness range (Figure 2).

(iii) Out of a large number of water variables examined, water hardness and certain associated water quality measurements, for example, nitrate, calcium, conductivity, carbonate hardness and silica give the strongest negative correlations with CVD mortality. The proportion of water supply derived from upland sources gives an equally strong but positive correlation. At this stage it can only be concluded that CVD is influenced by water hardness or by some factor closely associated with water hardness. This could be either a harmful factor in soft water or a protective factor in hard water.

These conclusions have been strengthened by work which has shown that in situations where the hardness of water supplies has changed there is a corresponding change in the CVD mortality rate at least for men[8]. The size of the water factor is small in comparison with other CVD risk factors e.g. heavy smoking doubles the risk of a CVD event while very soft water increases it by 10%.

Further research is proceeding and it has been concluded that the present situation does not warrant such drastic, unwelcome, costly and barely practical action as the hardening of soft water supplies. Some restriction will be placed on the level of central softening by the EC Directive and the manufacturers of domestic water softening equipment are being encouraged to provide a hard water tap to provide water for drinking and culinary purposes.

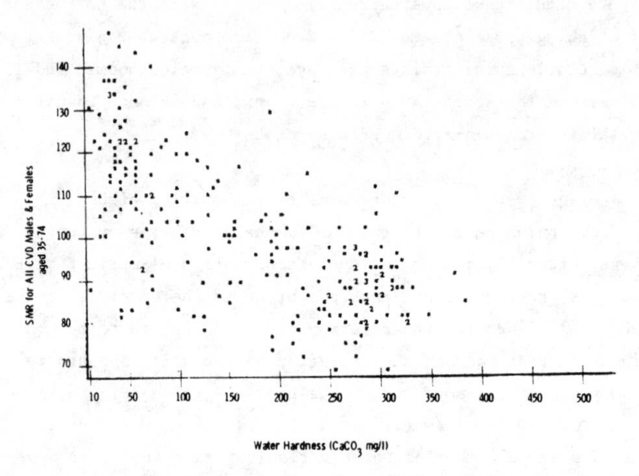

Fig. 1. SMRs* for CVD (male and female) plotted against water hardness

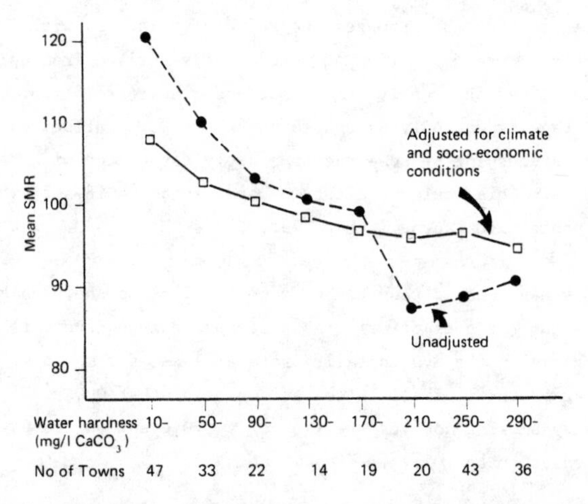

Fig. 2. Geometric means of SMR* (for all CVD males and females aged 35 to 74) for towns grouped in eight intervals of water hardness

* SMR (Standardised Mortality Ratio) = actual mortality rate expressed as percentage of rate expected on basis of age and sex profile of town i.e. 100 is 'normal'

ORGANIC COMPOUNDS AND WASTE-WATER RECYCLING

In England due to high population density and short rivers the levels of indirect waste water re-use are probably higher than in most other countries. (See Table 2.)

Table 2. Proportion of sewage effluent to total river flow at selected water supply abstraction points

River	Abstraction point	Proportion of effluent (%)	
		Average	Maximum*
Great Ouse	Foxcote	1.9	17.7
Great Ouse	Clapham	6.8	52.0
Great Ouse	Offord	12.2	58.3
Lee	New Gauge	16.5	81.4
Lee	Chingford	20.3	N/A
Thames	Buscot	5.6	51.1
Thames	Swinford	4.6	33.8
Thames	Sunnymeads	11.9	99.8
Thames	Staines	12.1	101.4
Thames	Surbiton	14.4	141.4

* Using river flow exceeded 95% of the time

This has led to concern that there could be a build up of non-biodegradable compounds that might affect health. Major interest in possible effects on cancer of gastro-intestinal and genito-urinary tracts.

Three types of study are being undertaken.

1. Water Quality Studies[9]. Different types of water supply (e.g. with varying degrees of re-use) are being analysed by GC-MS to see if there is evidence that supplies derived from sources containing a high proportion of sewage effluent contain more potentially hazardous substances. So far there is little evidence that this is so and a wide range of compounds including some known to be hazardous at higher concentrations (see Table 3) are found in many supplies at low concentration (usually < 1 μg/l). Unfortunately GC-MS can only be applied to the volatile and thermally stable components of water. Good methods for non-volatile organics are being urgently sought.

Table 3. Suspect and known carcinogens mutagens and promoters identified in drinking water

Compound	Frequency of occurrence (14 supplies)	Type of toxic agent
Acenaphthene	2	Mutagen
Benzene	13	Carcinogen
1-Bromobutane	1	Mutagen
Bromochloromethane	2	Mutagen
Bromodichloromethane	14	Mutagen
Bromoform	14	Mutagen
Carbon tetrachloride	6	Carcinogen
Chlorodibromomethane	14	Mutagen
Chloroform	14	Carcinogen
n-Decane	14	Promoter
Decanoic acid	3	Promoter
Decanol	1	Promoter
1,2-Dichloroethane	1	Mutagen
1,2-Dichloroethylene	2	Mutagen
Dibromomethane	7	Mutagen
1,4-Dioxane	1	Carcinogen
n-Dodecane	11	Promoter
Eicosane	2	Promoter
Fluoranthene	1	Mutagen
Hexachloroethane	1	Carcinogen
9-Methyl-fluorene	3	Mutagen
Octadecane	7	Promoter
Phenanthrene	7	Mutagen
Phenol	5	Promoter
Tetrachloroethylene	11	Carcinogen
1,1,2,2,-Tetrachloroethane	1	Carcinogen
Tetradecane	8	Promoter
Trichloroethylene	13	Carcinogen/ Mutagen
1,1,1-Trichloroethane	3	Mutagen
2,4,6-Trichlorophenol	1	Carcinogen
2,4,5-Trichlorophenol	1	Mutagen
Undecane	12	Promoter

2. Epidemiological studies. The health of populations drinking different types of water has been examined to see if any trends are evident. The most detailed study focused on the London boroughs[10]. It was essential to take account of socioeconomic factors which are known to have a influence on cancer incidence. When this was done no statistically significant relationships between cancer incidence and re-use could be found. A national study gave substantially similar results. It has to be recognised that because of the large number of factors that influence cancer incidence, epidemiological studies of this kind are rather insensitive.

3. Toxicological studies. One approach being used is to apply biological mutagenicity tests[11] to water and water extracts. A positive result in the Ames test for example indicates the probable presence of a potential carcinogen. The positive sample is then fractionated, the fractions retested and analysed in an attempt to identify the mutagen present. This work is in progress.

ORGANIC COMPOUNDS AND DISINFECTION

The discovery in 1974 that (a) those parts of New Orleans served with Mississippi derived water had higher cancer rates than those served with groundwater, (b) chloroform, a suspected carcinogen, and other trihalomethanes were present in the river derived supplies, provoked considerable research in the USA and elsewhere.

This has shown that

1. Chloroform and other trihalomethanes (THM) are formed when water containing 'natural' organics (e.g. humic substances) is chlorinated in the disinfection process[12]. The formation of THM is time dependent and continues in the distribution system.

2. The formation of THMs can be minimised by

   (a) removing as much organic material as possible prior to chlorination.

(b) minimising the level of chlorination consistent with efficient disinfection,

(c) using alternative disinfectants such as chlorine dioxide and ozone. (Caution: these may not form trihalomethanes, but presumably they form something else! Studies are going on.)

3. Numerous epidemiological studies have been undertaken[13] and although these tend to show a slight excess cancer mortality associated with the consumption of river derived water, the possibility of confounding factors cannot be ruled out and the trihalomethanes themselves have not positively been shown to be hazardous substances at the concentration at which they are present in drinking water.

4. There is tremendous difficulty in extrapolating from the available toxicological data for chloroform and indeed other carcinogens to obtain a 'safe level' for drinking water. At best a variation in the result over two orders of magnitude is possible. This is reflected in the limits set for total trihalomethanes (chloroform + bromodichloromethane + dibromochloromethane + bromoform) in different countries i.e. Germany 25 μg/l, Netherlands 75 μg/l, USA 100 μg/l, Canada 350 μg/l. Table 4 shows the results of a small survey of levels in Britain carried out in 1980[14].

Table 4 National total trihalomethane concentrations

	total trihalomethane concentration (μg/l)		
	Raw	Treated	Stored (7 day) Treated
Min	N.D*	N.D*	N.D*
Max	46	341	378
Mean	2	34	52

* N.D. means that the concentration of each of the four trihalomethanes in the sample was less than 1 μg/l.

In considering this situation in perspective it has to be remembered

(a) Water is by no means the only source of chloroform e.g. 'tingling fresh' toothpaste contains 0.5 - 2% while a few 5 ml spoonfuls of some well known cough medicines give an exposure equal to a lifetime of water consumption.

(b) Available evidence is that the additional risk due to trihalomethanes in water is small - equivalent to 1 or 2 extra cigarettes a day at most.

(c) Under no circumstances must disinfection efficiency ever be compromised.

Water Authorities in the UK are advised that the situation does not warrant massive expenditure but that where THM can be minimised without the commitment of considerable resources it would be prudent to do so.

Research is continuing to obtain better toxicological information on the trihalomethanes, to identify other byproducts from the treatment of water with chlorine and other disinfectants and to develop remedial water treatment processes.

OTHER WATER CONSTITUENTS

There are several other water constituents that are or have been an issue in relation to public health. The most important are

Fluoride	The subject of moral, political and scientific debate. Most scientists accept the evidence that fluoridation (1 mg/l) reduces the incidence of dental caries without harmful side effects. The moral issues (mass medication) are constantly debated and fluoridation is not widely practised, although it has DHSS backing.
Asbestos	Water is often transmitted through asbestos cement pipes but pick-up of fibres is small except under unusual conditions. Many water supplies contain asbestos fibres but toxicological studies have indicated that ingested as opposed to inhaled asbestos is not a problem.

Sodium	There is weak evidence of some health effects although this seems unlikely because of the small contribution of water sodium to dietary intake. Can contribute to hypernatraemia in babies but the main problem here is the use of unsuitable milk powders.
Polynuclear aromatic hydrocarbons (PAH)	This group contains many carcinogenic compounds. PAHs are found in most surface waters and are removed in treatment but there may be pick-up from coal-tar linings on pipes. These linings are no longer applied. Water derived PAH represents a minute fraction of the total intake from food and the atmosphere.

CONCLUSIONS

There is only very limited evidence that chemical constituents of drinking water are involved in health problems. In general, water components represent only a very small proportion of the dietary intake of chemicals. Any health effects of such materials are likely to result from long-term exposure and unless this leads to an extremely unusual disease the effect is likely to be very difficult to detect.

Investigations into the basis of the observed statistical relationship between water softness and mortality from cardiovascular disease are of the classical type in which a cause is sought for an observed effect. In much of the other work referred to it may be said that possible causes have been identified for which we are seeking possible effects. This might give rise to the criticism that resources are being wasted in the pursuit of non-existent problems.

The justification for this work is the strong evidence that exists, that environmental factors are important in several chronic diseases including cancer. The average consumer has little opportunity to exercise choice in relation to the water that he drinks and it is therefore important that we should assess the risks and benefits associated with water constituents which we may be able to control. The quantification of risks is no easy matter and the work described represents only the initial steps in the process. What is urgently needed is a valid technique for comparing these risks with others that are accepted by the public on a day-to-day basis. Thus we need to know how the risk associated with a certain concentration of a toxic chemical in drinking water compares with that involved in crossing a busy street, driving a car or taking part in recreational activities, e.g. winter sports, swimming, mountaineering. This approach has been discussed by Pochin[15].

Risk assessment is necessary to protect public money as well as public health. The enforcement of unnecessarily stringent water quality standards could involve expenditure of millions of pounds which might be used to improve public health more effectively in other ways.

REFERENCES

1. World Health Organization. European Standards for Drinking Water. Second Edition. WHO, Geneva, 1970.

2. World Health Organization. International Standards for Drinking Water. Third Edition. WHO, Geneva, 1971.

3. Council of the European Communities. Council Directive of 15 July 1980 relating to the quality of water intended for human consumption (80/778/EEC). Official Journal of the European Communities, No L229/11, 30 August, 1980.

4. Department of Health and Social Security. Lead and Health. The Report of a DHSS Working Party on Lead in the Environment. HMSO, London, 1980.

5. Department of the Environment. Lead in drinking water - a survey in Great Britain 1975-1976. Pollution Paper No. 12. HMSO, London, 1977.

6. Fraser, P. and Chilvers, C. Health aspects of nitrate in drinking water. Studies of Environmental Science 12, Water Supply and Health. Ed. H. van Lelyveld and B.C.J. Zoeteman. Elsevier, Oxford, 1981.

7. Pocock, S.J. *et al.* British Regional Heart Study: geographic variations in cardiovascular mortality, and the role of water quality. Br. Med. J., 1980, 280, 1243-1249.

8. Lacey, R.F. Changes in Water Hardness and Cardiovascular Death-Rates. Water Research Centre Technical Report, TR 171. Water Research Centre, 1981.

9. Fawell, J.K. and James, H.A. Problems of assessing the toxicological significance of organic micropollutants in drinking water. Report of Conference. Organic Micropollutants in Water, 13 March 1981. Ed. Hammerton, D. Institute of Biology, London, 1982.

10. Packham, R.F., Beresford, S.A.A. and Fielding, M. Health related studies of organic compounds in relation to reuse in the United Kingdom. The Science of the Total Environment (1981), 18, 167-186.

11. Forster, R. and Wilson, I. The application of mutagenicity testing to drinking water. J. Inst. Wat. Engrs Sci., 1981, 35, 259-274.

12. Tressell, R.R. and Umphres, M.D. The formation of trihalomethanes. J. Amer. Wat. Wks Assn.,. 1978, 70, 604-612.

13. Wilkins, J.R. *et al.* Organic chemical contaminants in drinking water and cancer. Am. J. Epidemiology, 1979, 110, 420-448.

14. Hyde, R.A. Conclusions from trihalomethane survey. Papers and proceedings of Water Research Centre seminar on Trihalomethanes in Water, 16-17 January 1980. Water Research Centre, 1980.

15. Pochin, E.E. The acceptance of risk. Brit. Med. Bull., (1975), 32, 184.

3
Aspects of the Chemistry and Analysis of Substances of Concern in the Water Cycle

By R. Perry
DEPARTMENT OF CIVIL ENGINEERING, IMPERIAL COLLEGE, LONDON SW7 2BU, U.K.

In recent years, concern has been shown about the presence in the water cycle of trace chemicals. These materials, which may have been discharged both from domestic and industrial sources, vary immensely in chemical character and include those that are non-biodegradable as well as those formed in both water and waste water treatment processes.

Difficulties are frequently encountered in equating health related effects to trace levels of these materials although recently the World Health Organisation have recommended more comprehensive guidelines for acceptable levels of these materials in drinking water.

In defining levels of exposure to substances of concern, problems are frequently experienced in the analytical chemistry involved as major difficulties exist in determining trace substances in such complex environmental samples as sewage sludges when compared to monitoring the same materials in pure water.

Accordingly, the purpose of this lecture will not be to review the vast list of materials that can be included in the term "Substances of Concern", but rather to examine two typical situations where considerable debate exists as to acceptable levels of exposure and where analytical problems are involved in determining with specificity and sensitivity the levels involved.

The Formation and Control of Trihalomethanes in Water Treatment Processes

The occurrence of trihalomethanes (THM) in water supplies was first demonstrated by Rook who associated their formation with the process of water chlorination. Although a wide range of halogenated compounds can be formed during chlorination, THM are usually considered as the four principal compounds formed, namely chloroform, bromodichloromethane, dibromochloromethane and bromoform.

Growing international concern over these materials, associated to some extent with the suggested carcinogenicity of chloroform, has led to much discussion about legislation with the incorporation of this group of compounds into water quality standards.

In 1979, the Environmental Protection Agnecy of the United States announced regulations limiting the permissible level of THM in potable water supplies to 100 $\mu g\ l^{-1}$ determined as a mean annual concentration. This measure has aroused a great deal of discussion with regard to both the cost of implementation and the scientific validity of the decision made. Implementation would involve the costly use of granular activated carbon in many water treatment works at a time when there is little evidence relating present levels of THM to health effects. Here the situation is further complicated by the balance between chlorinated and brominated materials in that almost nothing is known about the health related effects associated with the brominated materials.

Other countries are following the U.S. lead in legislation and suggested standards have included 350 $\mu g\ l^{-1}$ for THM in Canada with Germany and the Netherlands proposing standards of 25 and 75 $\mu g\ l^{-1}$ respectively.

It is evident that the controversy surrounding THM formation is complex scientifically and that legislation is being introduced on the basis of inadequate data. Decisions made purely on scientific grounds are rendered more difficult by the present toxicological tests which are clearly inadequate to resolve environmental issues where risk factors of 1 in 10^6 are the norm.

These low risk factors associated with THM formation in water supplies are of course in no way comparable with the risks that would be incurred should the process of water chlorination be terminated. On no account should the process of removal of trace organics resulting from the chlorination of water supplies jeopardize the present high standards of water disinfection.

THM Formation in Water

THM levels reported clearly reflect modes of practice adopted in water treatment and distribution in different countries as well as the quality of the source waters. Thus in Germany, for instance, where utilization of waters with total organic carbon (TOC) levels reduced to 2 $mg\ l^{-1}$ and chlorine dosing limited to less than 1 mg l^{-1}, THM levels are much lower than in the U.S.A. where prechlorinating procedures largely account for the higher levels recorded. Similarly, in the U.K. higher than average THM levels are associated with prechlorination and the utilization of some source waters with high TOC contents.

Clearly, the quality of the source water, whether this be a lowland surface water, an upland water or a groundwater, is important as THM formation is largely dependent upon the nature of the organic content of that water. Many precursors have been suggested in evaluating the mechanism of formation of THM but humic and fulvic materials present in natural waters are considered to be the most important. These, together with certain model compounds, have been evaluated in studying the effects of such treatment parameters as temperature, pH and chlorine dose on THM formation.

In lowland water re-use situations, ammonia levels also affect the level of THM formation and here too the level of bromide is significant in determining the balance of halogenated organic materials obtained.

Seasonal variations of THM concentrations are a function not only of water temperature but also relate to the time when the maximum concentration of precursor occurs. Here the formation of algal blooms in summer has also been related to THM formation.

Aqueous Chlorination Reactions

The complexities of water chlorination have been extensively studied and much attention has been paid to the disinfection of lowland river sources with high chlorine demand.

The diversity of aqueous chlorination reactions stems from the ability of the chlorine molecule to either hydrolyse or disproportionate rapidly with the following equilibrium reactions involved

$$Cl_2 + H_2O \rightarrow Cl_2(aq) \rightleftarrows HOCl + H^+ + Cl^- \quad (K = 1.5 - 4.0 \times 10^{-4}) \tag{1}$$

$$HOCl \rightleftarrows H^+ + OCl^- \quad (K = 1.6 - 3.2 \times 10^{-8}) \tag{2}$$

These equilibria are pH dependent but for most natural waters the buffering capacity ensures that hypochlorous acid (HOCl) and the hypochlorite ion (OCl^-) are the predominant species present, the proportion of each depending on pH.

It is considered that the major species involved in the chlorination reaction is the HOCl and the hypochlorous acidium ion as these are considerably more reactive

$$HOCl + H^+ \rightleftarrows H_2OCl^+ \qquad (3)$$

than the hypochlorite ion (OCl^-). The latter is, however, important in oxidation reactions.

For the THM formation therefore the balance on chlorination between substitution reactions involving initially the formation of either chlorine-carbon or chlorine-nitrogen bonds and general oxidation reactions is important. Addition reactions are relatively unimportant as these are generally much slower.

The haloform reaction involves a series of base-catalysed substitution reactions on an α-carbon to a carbonyl group followed by eventual hydrolysis to produce a trihalomethane. This is illustrated in Figure 1. For simple acetyl-containing compounds the slow rate-determining step is proton dissociation which is independent of halogen concentration but strongly pH dependent.

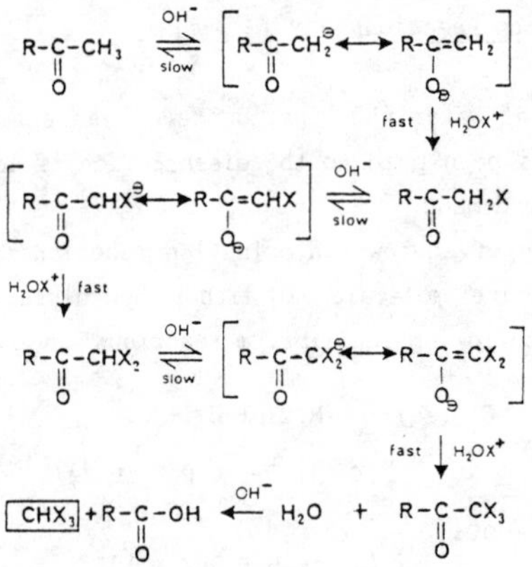

$$R\text{-}CO\text{-}CH_3 \xrightleftharpoons[\text{slow}]{OH^-} [R\text{-}CO\text{-}CH_2^{\ominus} \longleftrightarrow R\text{-}C(O^{\ominus})\text{=}CH_2] \xrightarrow[\text{fast}]{H_2OX^+} R\text{-}CO\text{-}CH_2X$$

$$R\text{-}CO\text{-}CH_2X \xrightleftharpoons[\text{slow}]{OH^-} [R\text{-}CO\text{-}CHX^{\ominus} \longleftrightarrow R\text{-}C(O^{\ominus})\text{=}CHX] \xrightarrow[\text{fast}]{H_2OX^+} R\text{-}CO\text{-}CHX_2$$

$$R\text{-}CO\text{-}CHX_2 \xrightleftharpoons[\text{slow}]{OH^-} [R\text{-}CO\text{-}CX_2^{\ominus} \longleftrightarrow R\text{-}C(O^{\ominus})\text{=}CX_2] \xrightarrow[\text{fast}]{H_2OX^+} R\text{-}CO\text{-}CX_3$$

$$R\text{-}CO\text{-}CX_3 + H_2O \xrightarrow{OH^-} \boxed{CHX_3} + R\text{-}CO\text{-}OH$$

Figure 1 Reaction mechanism of the haloform reaction

THM Precursors

Acetone was the first suggested precursor of THM in water chlorination processes. It was subsequently shown however that it would need to be present at mg l^{-1} concentrations to account for the chloroform levels monitored and that in addition the rate of chloroform production from acetone was far too slow to be of significance.

Yields of chloroform from humic materials have been reported as between 0.2 and 1.6% based upon available carbon whilst accounting for less than 5% of the total chlorine consumed.

Analytical Procedures

Specific analytical procedures have been developed allowing for evaluation of intermediates formed in the chlorination process as well as the dissolved THM. In this way the total potential of a water for THM formation can be assessed as being a measure of the dissolved THM and the THM associated intermediates that could subsequently break down in the distribution system. (Residual).

The procedure employed for the separation of the four trihalomethanes of concern involved a direct aqueous injection (DAI) chromotographic technique. Using an injection port temperature of 200°C, the intermediates formed in the chlorination reaction that might normally break down slowly in the distribution system, break down rapidly allowing both the dissolved THM and the residual THM to be passed to the chromosorb 102 chromatographic column for analysis (Total THM).

By first purging the sample with nitrogen dissolved THM can be removed and subsequent DAI analysis relates to the breakdown of the intermediates (Residual THM).

Dissolved THM levels, analogous to those obtained using conventional solvent extraction procedures or purge-and-trap methods, can be obtained by difference.

$$\text{TOTAL THM} = \text{RESIDUAL THM} + \text{DISSOLVED THM} \quad (4)$$

It is important to stress that the residual THM is due to the breakdown of chlorinated organic intermediates and it is not due to an accelerated reaction with any remaining free chlorine. This was ensured by the addition of a reducing agent to the sample prior to analysis.

Figure 2 illustrates the influence of pH on chloroform production from humic acid whilst Figure 3 shows the effect of temperature variation. At both low pH and low temperature the rate of dissolved THM formation is considerably reduced although the situation is complicated somewhat when residual THM is taken into account.

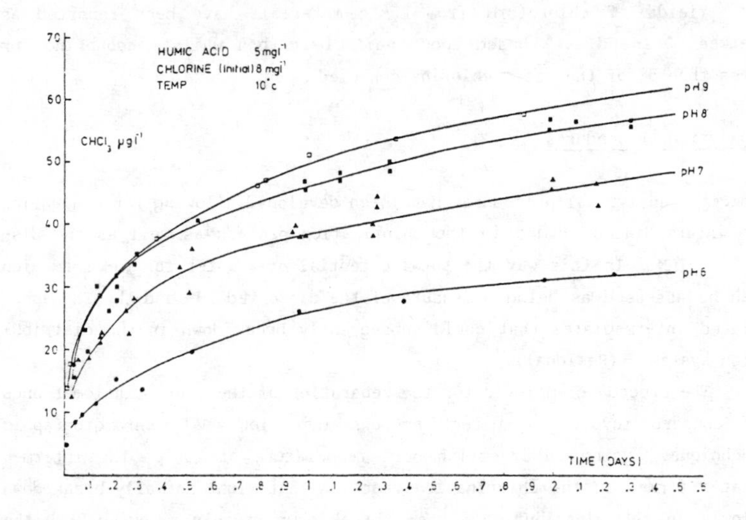

Figure 2 Effect of pH on dissolved chloroform formation

In all the work which has been carried out higher dissolved chloroform levels were obtained from humic acid (H.A.) as compared to fulvic acid (F.A.), with relative concentrations being

$$\frac{[CHCl_3]_{humic\ acid}}{[CHCl_3]_{fulvic\ acid}} \cong 1.7 \qquad (5)$$

This difference is undoubtedly related to the number of active sites of the respective materials.

Similar relationships have been found to hold with respect to pH and temperature for upland waters containing different humic materials to

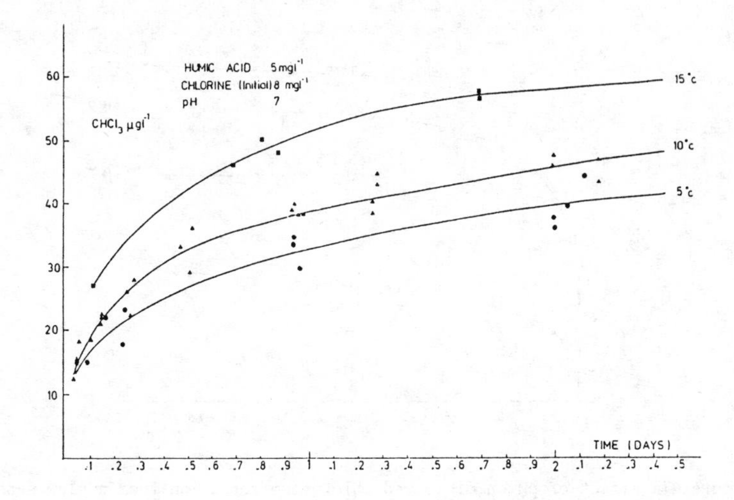

Figure 3 Effect of temperature on dissolved chloroform formation from humic acid at pH 7

those extracted from the Thames. As can be seen from Figure 4, which illustrates dissolved chloroform production from a slow sand filtered upland water, THM product is rapid initially and reaches a maximum within 12 hours at all pH values. It is clear from work that both the concentration and nature of the humic materials which is dependent on the source of the water determines the overall levels of trihalomethane formation.

Water rich in humic materials has a high chlorine demand and at low chlorine doses rapidly exhausts free chlorine. It might be expected under these circumstances that THM production would cease. This however is not the situation as considerable quantities of halogenated organic intermediates have been formed which subsequently break down hydrolytically to produce further THM.

The formation of these materials, which have been classified as residual THM, is also pH and temperature dependent. The balance between residual and dissolved chloroform formation after 2 days reactions from humic acid is given in Table 1.

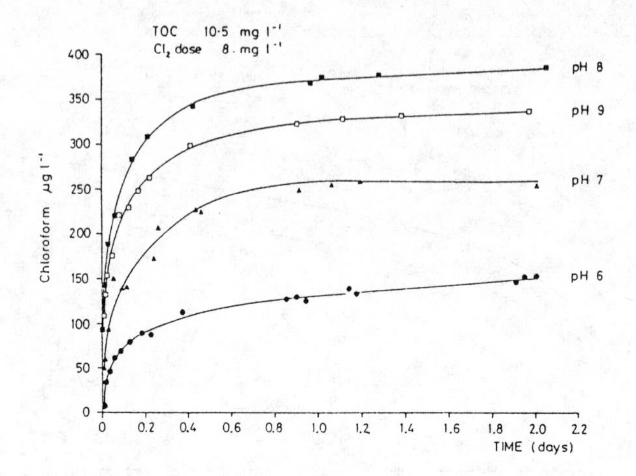

Figure 4 Effect of pH on dissolved chloroform formation from a slow sand filtered upland water

Table 1 Effect of pH on total, dissolved and residual chloroform concentration ($\mu g\ l^{-1}$) from humic acid after 2 days reaction at 10°C

pH	Total $CHCl_3$	Residual $CHCl_3$ (% of total)	Dissolved $CHCl_3$ (% of total)
6	67.5	36.5 (54)	32 (46)
7	87	42 (48)	45 (52)
8	92.5	37.5 (40)	55 (60)
9	79	20 (25)	59 (75)

Model Compounds and Mechanisms of Reaction

Research work utilizing model compounds has shed some light on possible mechanisms for THM formation. Many of the materials used for study relate in structure to fragments of humic materials and included in the groups of materials evaluated therefore have been a number of meta-dihydroxy-aromatic structures. Detailed studies have been carried out in order to assess the parameters that control chloroform formation from resorcinol and phlorizin as compared to humic materials. Although there is a striking similarity in the nature and behaviour of the intermediates produced from resorcinol and those produced from humic materials, the rate of formation and break down of the model compounds intermediates is significantly faster than that for the humic material itself.

Effect of Bromide on Trihalomethane Formation

Improved germicidal activity has been observed during chlorination in waters containing bromide. This has been attributed to the oxidation of bromide by chlorine producing the corresponding hypohalous acid.

$$HOCl + Br^- \rightleftarrows HOBr + Cl^- \qquad (6)$$

The hypohalous acid in acid conditions establishes the equilibrium

$$BrCl + H_2O \rightleftarrows HOBr + HCl \quad (K = 2.94\times10^5 \text{ at } 0°C) \qquad (7)$$

The presence of bromide, therefore, will give rise to not only chlorinated but also brominated compounds and, as hypobromous acid is formed in preference to hypochlorous acid, the degree of bromination will be dependent on their relative concentrations. Studies on the relative rates of reaction of bromine chloride and of bromine with a variety of aromatic hydrocarbons show that the reactivity of BrCl is, on average, greater than that of Br_2 by a factor of 20, but both yield only brominated compounds. Thus, bromine chloride acts solely as a brominating agent with aromatic compounds in electrophilic substitution reactions as hypobromous acid is the predominant electrophilic species in solution. In addition reaction BrCl is 100 times more reactive than Br_2.

Few data are presently available regarding bromide levels in ground and surface waters, nor has any quantitative study been carried out relating to the transfer of bromide into these waters where possible sources could include vehicle emissions, bromide present in rock salt used in deicing activities, fumigant used horticulturally and bromide impurities in the chlorine used for water disinfection.

The presence of bromide is known to increase the formation of bromine containing trihalomethanes and there is some evidence that the total trihalomethane yield is also increased as a consequence of the greater chemical reactivity of the hypobromous acid. A detailed study has shown that for an average bromide content of 90 µg l^{-1} THM formation on chlorination has produced 46% of the total trihalomethanes as chloroform and 54% as bromine containing trihalomethanes (39% bromodichloromethane and 15% chlorodibromomethane). By contrast analysis carried out on an unpolluted upland water with a bromide concentration of less than 10 µg l^{-1}, produced over 95% chloroform and less than 5% bromodichloromethane. Clearly, the higher the bromide concentration in the raw water the higher the yield of brominated trihalomethanes. Of comparable significance is the assessment in the bromide chlorine ratio dependent on the chlorine dose.

The relationships however are extremely complex and as yet little is known of the relative stability of the organo bromine intermediates. The balance between residual and dissolved components needs to be evaluated in relation to the initial bromide content of the source water.

A Water Treatment Plant Investigation

The change in the chloroform content of water through a water treatment plant has been studied (Table 2) (Figure 5).

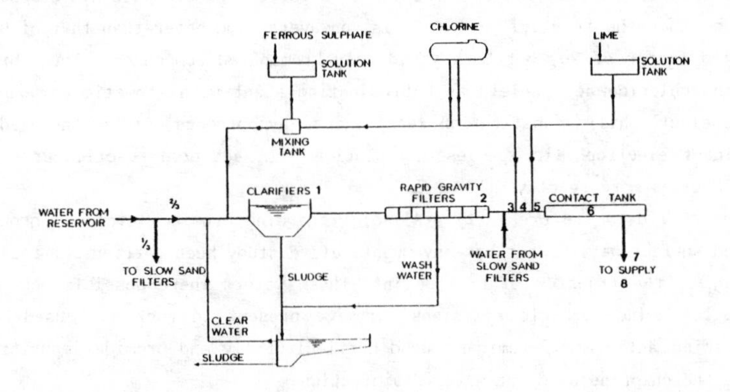

Figure 5 Water treatment plant flow diagram

Table 2 Chloroform and total chlorine concentrations at different stages of treatment and distribution

Sampling Site	Mean $CHCl_3$ µg l^{-1}			No. of Samples	Tot. Chlorine Residual Range (mg l^{-1})
	Tot.	Res.	Dis.		
1. Clarification	33	28	5	1	-
2. RGF	36	27	9	9	0.05-0.1
3. Mixed RGF/SSF (Prior to Cl_2)	28	23	5	3	N.D.
4. After chlorine	43	20	23	4	1.0-4.4
5. After lime	42	11	31	2	1.3-2.5
6. Contact tank (15 mins retention)	63	31	32	12	1.5-2.3
7. Final water (40 mins retention)	84	32	52	63	1.0-1.7
8. Distribution system					
(a) 3 hrs retention (estimated)	129	26	103	7	02.-0.8
(b) 4 hrs retention (estimated)	128	35	103	10	0.2-0.35

The chlorine used during the copperas treatment not only oxidized the ferrous sulphate to ferric sulphate but in addition reacted with the humic materials to produce chloroform. For effective coagulation the pH was maintained between pH 4.5 and pH 5 which was reflected in the high proportion of residual chloroform formed as measured in samples taken during clarification and after rapid gravity filtration.

After rapid gravity filtration the treated water (67%) was mixed with slow sand filtered water (33%) diluting the chloroform and removing the last traces of combined residual chlorine present in the rapid gravity filtered (RGF) water. The mixed RGF and slow sand filtered (SSF) water was subsequently chlorinated prior to lime addition. The total chlorine residual concentration at the sampling point immediately after chlorination varied considerably ranging between 1 mg l^{-1} and 4.4 mg l^{-1} as a result of incomplete mixing. After lime addition the dissolved chloroform concentration started to increase relative to the residual chloroform, consistent with the increase in pH. In the contact tank (after approximately 15 minutes contact) a significant increase in total chloroform formation occurred with stable chlorine concentration. However this was still only half the concentration found in the distribution system after an estimated 3 to 4 hours retention.

Control of THM Formation

Recommendations for THM control range from minor alterations in existing treatment practice to high cost alternatives in plant design which include the use of activated carbon for adsorption and air stripping. It is essential however that any proposed change in treatment practice should not lead to a deterioration in the microbiological quality of the finished water.

The methods for THM control can be classified as follows:

(a) Removal or reduction of precursors
(b) Change in treatment practice
(c) Removal from the finished water
(d) Use of alternative disinfection processes

The most effective method of reducing THM formation is to reduce the concentration of precursors prior to chlorination by either chemical oxidation of physico-chemical techniques. Unquestionably pre-chlorination leads to considerably elevated levels of THM and for this reason alternative oxidizing agents have been evaluated.

Chlorine dioxide treatment prior to chlorination was moderately unsuccessful in reducing THM precursors whereas ozonization, although more effective, has been shown to require a high dosing rate for significant precursor removal. Clearly precursor removal is strongly pH dependent in all these situations.

Such treatment undoubtedly produces organic byproducts which in themselves have unknown health effects. It is generally recommended therefore that ozonation be followed by filtration (biologically activated carbon) or coagulation.

Powdered activated carbon (PAC) has proved an effective method for reducing taste and odours in water supplies. However, for the removal of THM and their precursors it has been found uneconomical on account of the unrealistically high doses required for effective treatment.

Granular activated carbon (GAC) has proved more effective and is recommended by the EPA as the method most suitable for reducing THM. When fresh, the activated carbon will adsorb most trihalomethane precursors and produces a water with a low overall concentration of organic matter. Thus trihalomethane and hitherto unknown organic products are significantly reduced on chlorinating waters which have been filtered through granular activated carbon.

There are disadvantages however in that the activated carbon must be regenerated frequently so as to prevent breakthrough and maintain the high removal efficiency for organic matter. The frequency of regeneration is dependent on the organic load on the filters, and the type of precursors to be removed.

If used for the removal of THM in the finished waters the effectiveness of GAC is only maintained during the first two to three weeks of operation after which time breakthrough of the trihalomethanes and other volatile organic compounds takes place, resulting in increased running costs because of the frequent regeneration requirements.

Coagulation utilizing both iron (III) and aluminium (III) salts is a well-used physico-chemical method for removing substantial proportions of organic matter from water. Coagulation is not as effective as granular activated carbon in removing organic matter, particularly synthetic or non polar compounds. However, there is growing evidence to suggest that coagulation is selective in removing precursors that readily produce trihalomethanes dependent on the molecular weight and number of active sites per unit weight. The organic material remaining after clarification consists primarily of low molecular weight humic material which, as discussed, tends to produce lower chloroform yields compared to the high molecular weight, highly coloured, organic matter.

Other methods for removal of THM from finished waters include aeration which removes volatile organic materials. However this has not proved very effective, even when using very high air to water ratios. A further drawback is the continued formation in the distribution system resulting from the breakdown of non-purgeable residual trihalomethanes.

Clearly, therefore, limitation of THM formation can be achieved by changes to existing treatment practice. The results will largely depend on the type of treatment and on flexibility of the system to change without a deterioration of the prime requirements of the works for a safe supply of potable water.

Some Environmental Implications of a Suggested Detergent Builder

Eutrophication of lakes and rivers is a problem in certain countries; this condition manifests itself by an abundance of algal growth. It is generally accepted that the onset of eutrophication may be promoted by an increase in the phosphorus discharges to the water concerned as a result of human activity.

Experiments have demonstrated that phosphorus is one of the most important nutrients required for the production of algal blooms. However, the relationship between the concentration of dissolved phosphorus in a body of water and the potential for algal growth is difficult to establish.

A considerable portion of the phosphorus discharged to natural waters enters in sewage effluent, but significant contributions may also arise from agricultural "run-off". Estimates for the phosphorus contribution from "run-off" in the United States have been as high as 45% while an estimate of 30% has been made for the source in the United Kingdom. A further contribution is made by atmospheric deposition. The phosphorus present in sewage originates principally from human excreta and detergents. Although the relative contribution from these two fractions may vary, depending on the water consumption and extent of detergent usage, the detergent contribution to the phosphorus in sewage is generally accepted as being 50%.

Sodium tripolyphosphate (STPP) has been the main compound used as a detergent builder for the last twenty or thirty years and is the main source of phosphorus in detergents. A typical heavy-duty detergent could contain 13.3% of phosphorus (by weight) as P from STPP.

Table 3 Constituents of a typical household detergent

Surfactant (anionic, non ionic and soap)
Builder
Corrosion inhibitor (sodium silicate)
Soil suspension agent (carboxymethyl cellulose)
Optical brightener
Sequestering agent (EDTA)
Bleaching agent (sodium perborate or separate hypochlorite)

It has been suggested that eutrophication may be controlled by the reduction or removal of detergent phosphorus from sewage. This view has stimulated the search for an alternative detergent builder. Amongst the compounds considered, the sodium salt of nitrilotriacetic acid (NTA) has generated considerable interest. This material has been extensively tested in Canada, Sweden and Finland, while other countries have decided against the use of this compound awaiting further information about its toxicity and impact on the environment.

The biodegradability of NTA has been extensively studied utilizing a variety of aerobic systems, Figure 6. One activated sludge plant simulation closely resembles the official German apparatus for testing the biodegradability of detergents. The original design of Husmann has been slightly modified to include 3 l reactor and 2.1 l settler constructed of borosilicate glass to avoid lead contamination. Experiments were performed at 18°C±2°C.

This work clearly indicates NTA to be biodegradable.

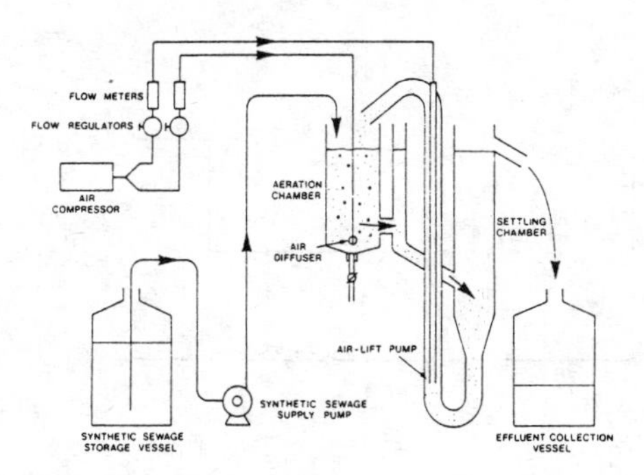

Figure 6 Modified official German laboratory apparatus for determination of the biodegradability of detergents.

The onset of biodegradation usually follows a period of acclimation, which varies from seven days up to several weeks. The acclimation period appears to be dependent on a number of factors including temperature, concentration of NTA, water hardness and the principal metal complexes of NTA which could be present.

Bacteria capable of utilising NTA as the sole source of both carbon and nitrogen exhibit optimum growth at 25°C. At temperatures above 30°C growth is significantly reduced, while at 2°C growth is extremely slow. However these organisms are capable of surviving temperatures below 0°C. In studies undertaken with a "fill and draw" system, where the influent concentration of NTA was 10 mg l^{-1}, biodegradation of 80% to 85% of the NTA was observed at 25°C, while only 25% was biodegraded at 5°C. Experiments at a full-scale sewage treatment plant indicated that more than 95% of the NTA was removed during the summer months but the average removal during the winter months dropped to approximately 50%.

It has been estimated that if STPP were entirely replaced by NTA as the principal detergent builder, concentrations of NTA in domestic sewage would be in the range of 10 to 20 mg l^{-1}.

With an influent concentration of 8 mg l^{-1} in a full-scale activated sludge treatment 90% biodegradation of NTA was achieved when the concentration was doubled although time was required for adjustement, Figure 7.

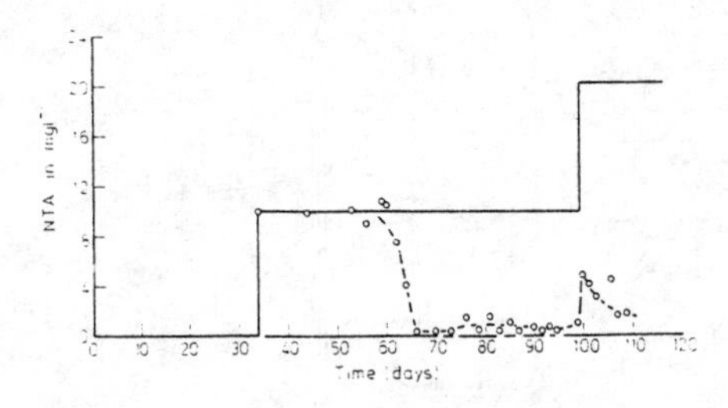

Figure 7 The variation in the influent and effluent concentration of NTA in a hard water activated sludge unit

Since NTA is a strong chelating agent, its incomplete biodegradation might be expected to have adverse effects upon heavy metal removal in sewage treatment processes and this has been shown to be the case.

The mobilization of heavy metals by NTA from river sediments and lake sediments has also been observed.

Table 4 Average heavy metal concentrations in the influent (I), effluent (E) and mixed liquor (ML) of activated sludge units supplied with synthetic sewage to which either nitrilotriacetic acid (NTA) or sodium tripolyphosphate (STPP) was added

Metal	Sample	Heavy Metal Concentration mg l^{-1} NTA	STPP
Cd	I	0.0120	0.0100
	E	0.0080	0.0045
	ML	0.13	>0.12
Cr	I	0.120	0.090
	E	0.021	0.008
	ML	3.0	>1.7
Cu	I	0.150	0.100
	E	0.140	0.082
	ML	1.7	0.35
Ni	I	0.15	0.11
	E	0.14	0.10
	ML	0.48	0.34
Pb	I	0.17	0.050
	E	0.044	0.004
	ML	5.0	>0.5
Zn	I	0.54	0.60
	E	0.44	0.09
	ML	4.9	>6.0
Average daily sludge wastage (ml)		635	737

Similar problems occur with the anaerobic degradation of NTA where again aclimation to biodegradation is the important factor, Figure 8.

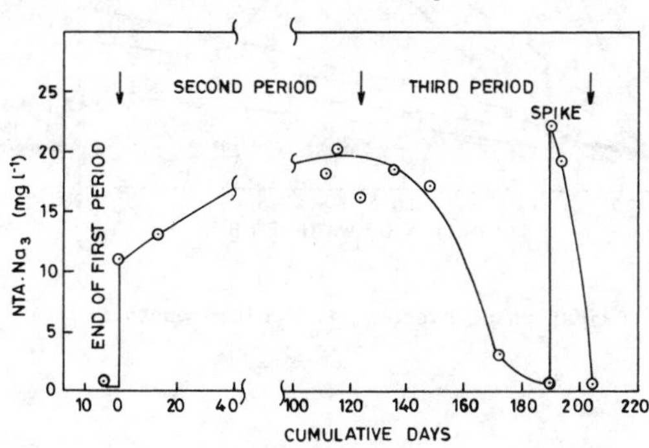

Figure 8 NTA.Na_3 concentration of digester, day 10 to 205.

Although some relationships have been demonstrated between biodegradation rate and chemical structure (Figure 9) such studies are complex and difficult to standardize.

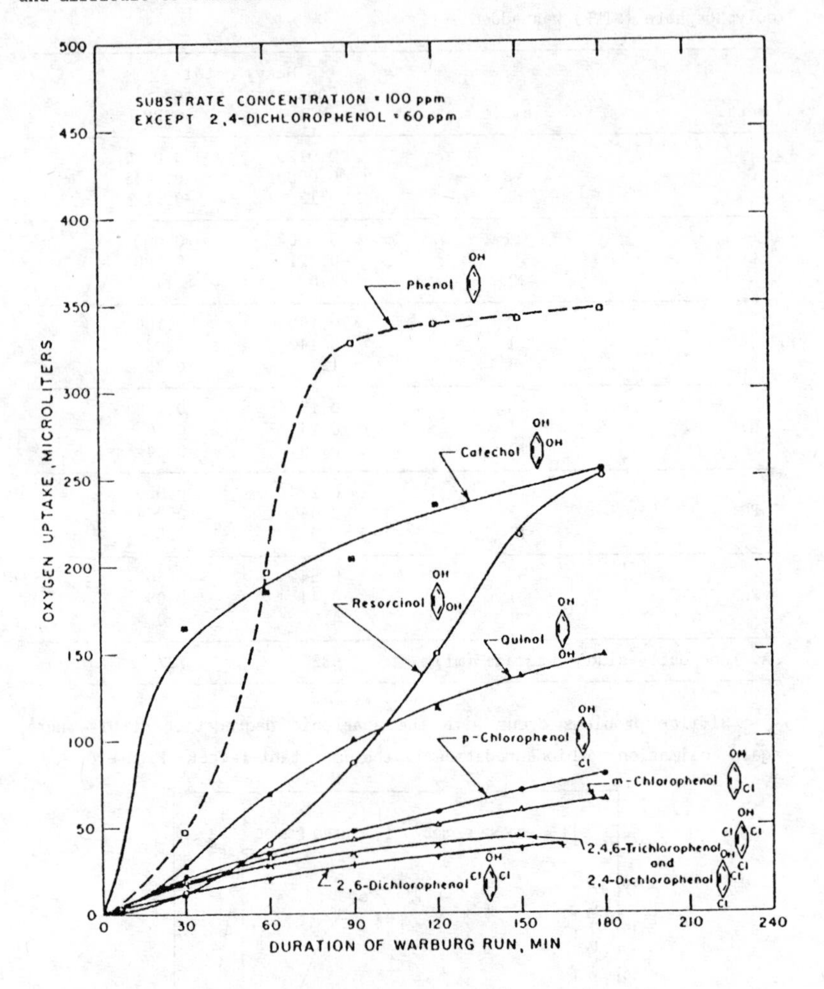

Figure 9 Oxidation of hydroxy- and chloro-phenols.

4
The Role of Wastewater Treatment Processes in the Removal of Toxic Pollutants

By A. James
DEPARTMENT OF CIVIL ENGINEERING, UNIVERSITY OF NEWCASTLE UPON TYNE, NEWCASTLE UPON TYNE NE1 7RU, U.K.

Introduction

Toxicity has been defined as the ability of a substance to produce injury once it reaches a suitable site in or on an organism. This definition would include the effects of organic matter and solids but the following discussion is restricted to substances which exert some chemical poisonous action.

Even within this restricted definition there are a large number of organic and inorganic substances which may interfere with aquatic organisms. Various methods of physical, chemical and biological treatment have been devised for treating wastes to limit the discharge of these substances to the aquatic environment.

The following notes consider in admixture:

a. Sources and types of toxic wastes
b. Pre-treatment on site
c. Treatment in admixture with domestic wastes
d. Two particular examples - tannery wastes or metal processing wastes

Sources and Types of Toxic Wastes

Domestic wastes contain ammoniacal nitrogen at levels up to 50-60 mg l^{-1} and when septic may contain sulphide at levels up to 50 mg l^{-1}, both of which can cause damage to aquatic life if discharged. Agricultural wastes may also contain even higher levels of both ammonia and sulphide plus a range of toxic pesticides and herbicides and the drainage from silage may be toxic due to low pH values. But the majority of toxic wastes are associated with industry. Here the range of possibilities is too wide to catalogue comprehensively but Table 1 gives some indication of the main types of toxic industrial waste and the toxins that they contain.

TABLE 1 Sources of some common toxins

Toxin	Sources
Acids - mainly inorganic but some organic causing pH < 6	Acid Manufacture Battery Manufacture Chemical Industry Steel Industry
Alkalis - causing pH > 9	Brewery wastes Food Industries Chemical Industry Textile Manufacture
Antibiotics	Pharmaceutical Industry
Ammoniacal nitrogen	Coke Manufacture Fertiliser Manufacture Rubber Industry
Chromium - mainly hexavalent but also less toxic trivalent form	Metal processing Tanneries
Cyanide	Coke production Metal Plating
Detergents - mainly anionic but some cationic	Detergent Manufacture Textile Manufacture Laundries Food Industry
Herbicides and Pesticides - mostly chlorinated hydrocarbons	Chemical Industry
Metals - mainly copper, cadmium, cobalt, lead, nickel, mercury and zinc	Metal processing and plating Chemical Industry
Phenols	Coke Production Oil Refining Wood Preserving
Solvents - mostly benzene, acetone, carbon tetrachloride, alcohols	Chemical Industry Pharmaceuticals

Pre-Treatment of Toxic Wastes

Pre-treatment of industrial wastes prior to discharge to a sewer is widely practised. In some cases this is to comply with the restrictions imposed by Water Authorities and Local Authorities on wastes entering a sewer. Typical consent conditions are given in Table 2.

TABLE 2. Typical Consent Conditions for Discharge to Sewers

Parameter	Consent Condition
Maximum Temperature	40-45°C
pH	6-10
Substances producing inflammable vapours	Nil
Cyanide concentration	$\ngtr$ 5-10 mg l^{-1}
Sulphide concentration	$\ngtr$ 1 mg l^{-1}
Soluble sulphates	$\ngtr$ 1250 mg l^{-1}
Synthetic detergents	$\ngtr$ 30 mg l^{-1}
Free chlorine	$\ngtr$ 100 mg l^{-1}
Mercury	$\ngtr$ 0.1 mg l^{-1}
Cadmium	$\ngtr$ 2 mg l^{-1}
Chromium	$\ngtr$ 5 mg l^{-1}
Lead	$\ngtr$ 5 mg l^{-1}
Zinc	$\ngtr$ 10 mg l^{-1}
Copper	$\ngtr$ 5 mg l^{-1}
Zinc equivalent (Zn + Cd + 2Cu + 8 Ni)	$\ngtr$ 35 mg l^{-1}
Total non-ferrous metal	$\ngtr$ 30 mg l^{-1}
Total soluble non-ferrous metal	$\ngtr$ 10 mg l^{-1}

There are also a large number of specific toxic substances whose discharge to sewers is controlled.

The main advantages of treatment on site are the possibilities of recovering heat or specific substances in an uncontaminated condition and the economies which might result from treatment at higher temperatures and/or higher concentrations. Where the toxic materials are organic in nature there is often some difficulty in treatment due to inhibition of bacterial growth. It is often easier and cheaper to develop the necessary bacterial flora in an on-site treatment plant.

This to some extent depends upon the concentration and toxicity of the substances concerned. In some cases dilution of the waste by admixture with sewage reduces the toxic inhibition making it preferable to treat the industrial waste and sewage together. Also many industrial wastes are deficient in some nutrient such as nitrogen or phosphorus. The desirable ratio of BOD:N:P is 100:5:1 and the ratio in domestic wastes is commonly 100:18:2.5 so that deficiencies in industrial wastes can be balanced.

There are other considerations in deciding for or against pre-treatment such as

a. Availability of space - the site may be too restricted or land may be too valuable to be used for a treatment plant.
b. Availability of expertise - the company may not wish to get involved in effluent treatment.
c. Sludge and/or odour production may create a nuisance.

Even where it is decided not to carry out pre-treatment of the toxic waste by chemical or biological methods it is often useful to install devices to improve the effluent quality by simple physical means. This includes some form of screening coarse or fine to reduce solids. Also some form of balancing to reduce variations in concentration, flow, pH etc, and some traps to prevent the escape of oil and grease and some grit arrestors.

Every attempt should be made to minimise the quantity of material discharged through good housekeeping. This can take the form of any or all of the following techniques:

i. Extending the life of process solutions by filtration, topping up, adsorption, etc.
ii. Altering the production process to use less toxic compounds, e.g. substitution of copper pyrophosphate for copper cyanide in electroplating solutions.
iii. Dry cleaning prior to wash-down can remove a large proportion of the pollutant in solid form.
iv. Evaporation of strong organic liquors can often produce a burnable product.
v. Minimising and segregating any flows which contain toxic materials. In some cases it is necessary to separate wastes for safety reasons, e.g. cyanides or sulphides and acid wastes, trichlorethylene and alkaline wastes. In other cases it may be desirable to segregate for treatment reasons. However segregation can be very expensive.

Having minimised so far as possible the types, quantities and concentrations of any toxic wastes it may still be necessary to treat them prior to discharge either to a sewer or a water course. The processes which are used may be classified as physical, chemical and biological. The physical processes are summarised in Table 3.

TABLE 3. Physical methods of pre-treatment

Process	Aim	Examples
Screening	Removal of coarse solids	Vegetable canneries, Paper Mills
Centrifuging	Concentration of solids	Sludge dewatering in chemical industry
Filtration	Concentration of fine solids	Final polishing and sludge dewatering in chemical and metal processing
Sedimentation	Removal of settleable solids	Separation of inorganic solids in ore extraction, coal and clay production
Flotation	Removal of low specific gravity solids and liquids	Separation of oil, grease and solids in chemical and food industry
Freezing	Concentration of liquids and sludges	Recovery of pickle liquor and non-ferrous metals
Solvent Extraction	Recovery of valuable materials	Coal Carbonizing and Plastics manufacture
Ion Exchange	Separation & Concentration	Metal processing
Reverse Osmosis	Separation of dissolved solids	Desalination of process and wash water
Adsorption	Concentration and removal of trace impurities	Pesticide manufacture, dyestuffs removal

Where the toxic wastes contain or are composed of organic materials it may also be necessary to provide some biological treatment especially if the effluent is to be discharged directly into a watercourse. Many different types of process are used but the following are the most popular.

a. High-rate filtration using plastic media with very high rates of recirculation.

b. Activated sludge using contact stabilization.

Like all biological processes these can suffer from toxicity problems especially where the concentration of toxin is not constant. In general terms it is easier for bacteria and other microorganisms to adapt to toxic substances than for organisms like worms, fly larvae etc. For this reason conventional percolating filters have not proved successful - the lack of grazing fauna has led to persistent ponding.

Due to a combination of high organic strength and inhibition from toxic substances it is unusual to obtain complete treatment of toxic industrial wastes by conventional primary and secondary treatment. The effluent from high-rate filters often has a BOD and COD similar to settled sewage and is either suitable for discharge to a sewer or for further biological treatment on site.

Chemical treatment of industrial wastes may be used in addition to and to some extent in place of biological treatment. The aims are somewhat different since biological treatment is mainly a way of oxidising organic matter or a way of converting it into a settleable form. Chemical treatment can also provide oxidation through chlorine, ozone, etc., but this is used only for oxidising particular compounds like cyanide since it is expensive and liable to lead to the production of undesirable chlorinated organics. It is mainly used for pH correction and improving the removal of solids. The commonest chemicals in use are shown in Table 4.

TABLE 4. Chemicals used in industrial waste treatment

Chemical	Purpose
Calcium hydroxide	pH adjustment, precipitation of metals and assisting sedimentation
Sodium hydroxide	Used mainly for pH adjustment in place of lime
Sodium carbonate	pH adjustment and precipitation of metals with soluble hydroxide
Carbon dioxide	pH adjustment
Aluminium sulphate	Solids separation
Ferrous sulphate	Solids separation
Chlorine	Oxidation

Primary and Secondary Treatment

Provided that the pre-treatment of toxic industrial wastes is successful then no difficulties should be encountered in subsequent treatment. However no pre-treatment system is perfect and malfunction will occasionally occur mostly due to variations in the manufacturing process. As a result toxic material together with possible overload of organics and solids may be passed on to the subsequent treatment stages.

The effect of toxic materials on primary sedimentation is insignificant, since this is a purely physical process of sedimentation and flocculation. However the effect of the primary sedimentation on toxic wastes can be very important. Toxic materials in suspension such as particulate metals are effectively removed. Also the flocculant material has a great capacity for adsorption and removes the majority of dissolved metals, pesticides and other toxic organics. In one respect this is beneficial since it renders the waste material less inhibitory for biological treatment but it selectively concentrates the toxins in the sludge and may give rise to problems in digestion and in sludge disposal. Some indication of the removal of metals during primary treatment is given in Table 5.

TABLE 5. Amounts of heavy metal ions removed from sewage by sludges

Heavy metal ion	Primary sedimentation		Percolating filter treatment		Activated-sludge process	
	Metal concentration in crude sewage (mg/l)	Proportion removed by treatment (per cent)	Metal concentration in settled sewage (mg/l)	Proportion removed by treatment (per cent)	Metal concentration in settled sewage (mg/l)	Proportion removed by treatment (per cent)
Copper					0.4	54
Copper	Up to 0.8	45	Up to 0.44	20	Up to 0.44	80
Copper					0.4–25	50–79
Copper	Up to 5	12				
Copper					28	90–93
Dichromate	(as Cr) Up to 1.2	28	(as Cr) Up to 0.86	32	(as Cr) Up to 0.86	67–70
Dichromate					4.0	6.3
Dichromate					0.5–2	ca 100
Dichromate					5	50
Dichromate					50	10
Iron (Ferric)	3–9	40	1.8–5.4	Nil	1.8–5.4	80
Lead	0.3–0.9	40	0.18–0.54	30	0.18–0.54	90
Nickel	0.1–0.3	20	0.08–10	40	0.08–0.24	30
Nickel					2.0	31
Nickel					2.5–10	30
Zinc	0.7–1.6	40	0.4–1.0	30	0.4–1.0	60
Zinc					2.5	90
Zinc					2.5	95
Zinc						
Zinc	Up to 5	12			7.5	100
Zinc					15	78
Zinc					20	74

The key to successful secondary treatment of wastes containing toxic materials is the adaptation of the microorganisms to the presence of the toxins. Bacteria and to a lesser extent protozoa show considerable ability to acclimatise to the presence of toxic substances and a great adaptability in degrading new synthetic organics. Metazoa are rather less adaptable and for this reason activated sludge is generally better than percolating filters for treating toxic wastes. Experience with treating toxic industrial wastes on percolating filters has shown frequent ponding problems due to a lack of activity by the grazing fauna. High-rate filters which utilise hydraulic scouring for film control have been used successfully and the high recirculation ratios help to dilute any incoming toxins. This helps to overcome the other disadvantage of filters which is due to the plug-flow nature of the process. Any shock loads of toxin are not as readily diluted as in a completely mixed reactor.

The activated sludge process is generally preferred for dealing with wastes containing an admixture of toxic materials. In particular the completely mixed version gives immediate dilution of any shock loads. The dangers with activated sludge are that:

a. The toxin may reach concentration which inhibits enzyme activity.
b. Some toxins also affect the bacterial surface and therefore affect settleability.

Nitrification is particularly sensitive to these problems.

It is therefore important in treating a toxic waste by either biological process that a microbial population is developed which is acclimatised to the presence of the toxin and in the case of degradable toxins contains sufficient numbers of organisms which can metabolise the toxins. These twin aspects of acclimatisation require great care in the start-up operation and may need a period of several months before successful operation is achieved.

Sludge Treatment and Disposal

Toxins which remain in solution during primary and secondary sedimentation do not appear in the sludges and thus cause no further difficulties. Toxicity in digestion may also occur due to soluble toxins in the treatment of industrial wastes by anaerobic methods. But the main problems are due to toxins which are in settleable form or are readily adsorbed, are selectively concentrated in the sludge and give rise to difficulties in digestion (and in subsequent disposal). The classes of toxins involved are mainly:

a. Metals
b. Chlorinated hydrocarbons
c. Organic solvents
d. Detergents

Some indication of toxic levels is given in Table 6.

TABLE 6. Toxic effects in anaerobic treatment

Toxin	Inhibitory concentration	
	In Sewage	In Sludge
Chromium	-	2
Cadmium	2	2
Copper	1.5	
Iron	10	
Lead	100	
Nickel	80	
Zinc	50	
Detergent		2% of Suspended Solids
Benzene		50-200
Carbon tetrachloride		10
Chloroform		0.1
Dichlorophen	1	
γ-BHC		48
Toluene		430-860

all concentrations in mg l^{-1}

Industrial Waste Treatment

The foregoing notes have dealt in general with the problems of treating toxic wastes but it is not possible to explain the complexities in so general a discussion. The following notes therefore describe in more detail two examples of industries that produce toxic wastes and the methods used in treating them.

Tannery Wastes

The waste is complex arising from the following operations:

a. Soaking - soil, dung, blood etc.
b. Dehairing - sodium sulphide and fibre

c. Liming - calcium hydroxide and protein

d. Deliming and bating - ammonium chloride and sodium bisulphite or boric, lactic and sulphuric acids + proteolytic enzymes + protein

e. Pickling - sulphuric acid, sodium chloride + protein

f. Tanning - Trivalent chromium and other neutral salts
- Catechols and pyrogallols
- Borax
- Alkali

g. Dyeing - Weak acid + dyes

The quality of wastes generated varies from 35-150 l/kg hides. The quality of effluent is also very variable due to batch process.

Parameter	Range
SS	0- 8,500
BOD	100-10,000
S^{2-}	0- 500
pH	3.0-11.0

These variations can be considerably reduced by balancing. Effluent from tanning may be disposed of in one of four ways.

a. Discharge to sewer
b. Discharge to river or estuary
c. Discharge to sea
d. Discharge underground

The discharge of tannery waste to a river or estuary is usually subject to severe constraints on solids, BOD, toxins, pH, etc. Only a minority of tanneries have access to the sea and though underground disposal is popular in USA it is not favourably regarded in UK.

Whether effluents are being discharged to sewers or to rivers the form of the pre-treatment is similar. In cases where space is very limited a mechanically brushed perforated screen is all that is installed but wherever possible a balancing tank should be added. The latter will reduce the need for pH correction, reduce the load of BOD and SS and enable any treatment units to operate continuously.

If the effluent is to be discharged to sewers some reduction in sulphide and metals may be necessary. Removal of specific contaminants such as chromium sulphide may be desirable but the isolation of drainage lines usually make this too expensive. It is however usually possible to separate the sulphide-containing liquors (dehairing and liming liquors). The aim is to reduce sulphides from concentrations of up to 2500 mg l^{-1} to a level of about 10 mg l^{-1}. Three methods of sulphide removal have been used:

a. Precipitation by copperas ($Fe\ SO_4$)
b. Oxidation by chlorine
c. Oxidation by dissolved air in the presence of a catalyst ($Mn\ SO_4$)

The relative economics of these processes favour the aeration technique. Chromium is less of a problem than sulphide due to the very poor solubility of trivalent chromium at pH values over 6. Also chromium in the spent liquors has been reduced in recent years by improvements in leather technology. Nevertheless at large tanneries recovery of chromium is economically viable.

The aim of the treatment is to reduce the chromium concentration below 20 mg l^{-1} which is the standard imposed by many Water Authorities for discharge. The tanning liquor may be treated with sodium hydroxide, sodium carbonate or calcium hydroxide. There is a danger with the sodium salts that at high pH values some of the chromium may redissolve. But where it is intended to recover the chromium for re-use they are preferred to the use of lime since the presence of calcium salts is undesirable. Recovery of chromium can be achieved by removing the precipated hydroxide in a filter press, dewatering by centrifuging and redissolving in sulphuric acid.

Pre-treatment of leather dressing wastes is often carried out prior to discharge to sewer for, although relatively small in quantity, they are a very high polluting load. Substantial reductions in strength may be obtained by precipitation with lime together with iron or aluminium salts. The process requires careful control to minimise the sludge production, which can be up to 80% of the volume of waste treated.

Where tannery effluents have been discharged to sewers there have been some reports of difficulties at the municipal treatment plants. These difficulties seem to have arisen mainly because of the increase in organic concentration rather than any toxic effect. Undiluted tannery wastes may have BOD values around 2000 mg l^{-1} and therefore requires a minimum of 3-4 times dilution to avoid problems of oxygen transfer (care is needed in the use of COD data since the ratio of COD:BOD is often twice that found in raw sewage). There have been suggestions that the sulphide present in tannery waste may have an adverse effect on the biological treatment in admixture with sewage, but this is unlikely to be serious at concentrations of < 10 mg l^{-1}. Higher concentrations around 25 mg l^{-1} are tolerated by the activated sludge process and up to 50 mg l^{-1} for short periods. Percolating filters are even more resistant and concentrations up to 100 mg l^{-1} of sulphide can be treated successfully. The severe restrictions imposed on discharges of sulphide are mainly to protect the health of workers in the sewer and the fabric of the sewer and are not aimed primarily at protecting the treatment plant. Metal toxicity from tannery wastes has not been found to be a problem. Most of the metal is in the form of trivalent chromium and around 80% of this is removed during primary sedimentation. The resulting sludges do not give rise to difficulties in digestion provided that the retention time exceeds 21 days.

Experience of treating tannery waste on its own has confirmed the general rule that it is better treated along with domestic waste. Anaerobic treatment has not been too successful with moderate BOD removals and some difficulty in treating the resultant liquor.

Activated sludge appears to have good potential for treating tannery waste although this has not been exploited. Loading rates of 0.5-1.0 kg BOD per kg MLSS per day can be used with BOD removals $\geq$ 90%. In particular the oxidation ditch seems an attractive form of the activated sludge process for treating tannery wastes.

Experience with biological filters has been somewhat mixed. It has proved to treat effluent on a conventional stone filter using loadings around 0.12 kg BOD per m^3 per day with removal rates of over 95%. High-rate filtration using plastic media is likewise possible but the economics of this are not attractive when compared with chemical treatment.

Metal-Processing Wastes

Discharge of metals to the aquatic environment has been a major cause of concern and the treatment of these wastes has consequently attracted considerable attention. Wastes containing metals may arise from a variety of industrial and agricultural operations including tanneries, paint manufacture, battery manufacture, pig wastes, etc. but the main source is from metal processing. The wastes from metal processing may be classified as follows:

a. Mining - ore production and washing - also contain inert SS
b. Ore Processing - smelting, refining, quenching, gas, scrubbing, etc. - also contain sulphides, ammonia and organics
c. Machining - metal particles from machining usually mixed with lubricants
d. Degreasing - metals mostly in solution with cyanides, alkalis and solvents
e. Pickling - acids with metals and metallic oxides in solution
f. Dipping - alkalis with sodium carbonate, dichromate etc. plus metals
g. Polishing - particles of metals and abrasives together
h. Electrochemical or chemical brightening & smoothing - acids mainly sulphuric, phosphoric, chromic and nitric with metals in solution
i. Cleaning - hot alkalis with detergents, cyanides and dilute acids plus metals in solution
j. Plating - acids, cyanides, chromium salts, pyrophosphates, sulphamates and fluoroborates plus metals in solution
k. Anodizing - chromium, cobalt, nickel and manganese in solution

The sources of wastes in metal processing are numerous and also extremely variable both in quantity and quality.

Metals in the wastes occur in forms ranging from large particles of pure metal in suspension to metallic ions and complexes in solution. The most appropriate method of treatment depends upon the form of the metal, its concentration, pH, other constituents of the waste and the desired effluent standard. The technique most commonly employed in treating metal processing wastes is precipitation using pH adjustment. The optimum pH for precipitation varies depending on the particular metal and where several metals are involved a compromise pH is used. A typical value is in the range 8.0-9.0. With amphoteric metals, notably zinc, care must be taken to avoid too high a pH to prevent the formation of zincates. It should also be appreciated that other constituents of the waste e.g. ammonia can significantly affect the solubility of the metal hydroxides and it is therefore not possible to predict accurately the level of residual metal in the treated effluent.

Whilst the hydroxide precipitation method is satisfactory for most metals encountered in effluents both hexavalent chromium and lead are not precipitated in this way. Hexavalent chromium is present in wastes from metal plating and must first be reduced to the trivalent form before treatment with lime or caustic soda. The reducing agents commonly used are sodium bisulphite, sulphur dioxide and occasionally ferrous sulphate. The reduction is carried out under acid conditions and subsequent addition of alkali precipitates trivalent chromium hydroxide.

In the case of lead, the hydrated oxide formed when lime or caustic soda is added to the lead waste has an appreciable solubility and the resulting effluent after removal of solids would normally be unsatisfactory for discharge to sewer or watercourse. However, basic lead carbonate has a very low solubility and therefore sodium carbonate can be used in place of lime as the precipitating agent. Like zinc, lead is amphoteric and redissolves as plumbate at high pH and so careful pH control must be exercised.

A particular type of precipitation system used in the metal plating industry is known as the Integrated Method of Treatment. The principal feature of this system is that the rinsing stage immediately after the metal plating stage is a chemical rinse which precipitates the metal from the liquid around the article being plated. A further water rinse is then required to wash off the treatment chemical. In the case of nickel plating the chemical rinse would contain sodium carbonate to precipitate nickel carbonate, whilst with chromium a prior stage to effect reduction from hexavalent to trivalent form would be required. The integrated system has the advantages that water re-use can be readily practised and that the metals are not precipitated in a mixture and so that also can be recovered. However, it is sometimes difficult to adapt the system to existing plating lines since it necessitates the placement of an extra tank in the line.

Once the metals have been precipitated from solution, liquid and solid phases must be separated. The traditional method for this stage of treatment is settlement in either a circular or rectangular tank. In small installations where the effluent flow is less than say 25 m^3/day, it is convenient to carry out the effluent treatment on a batch basis and to allow settlement to take place in the same tank as that used for reaction. For larger installations, a continuous flow system is required. The size of the tank depends on both the maximum effluent flow-rate and on the configuration adopted for the tank. The most common type of settlement vessel is of the vertical upward flow pattern having a central feed well, a peripheral collection launder and a sludge cone at the bottom. Clarification of the effluent can be enhanced by the use of flocculating agents. Obviously the size and mode of operation of the precipitation system significantly affects quality of the effluent but typical figures for a well-designed, efficiently operated, settlement system for metal hydroxide precipitates would be in the range 10-30 mg/l suspended solids.

Where space is at a premium, a compact settling system utilising parallel tilted plates or tubes can be used to perform the separation stage.

There are two factors which make this system efficient in terms of ground area used. These are:

a. the distance through which a settling particle has to fall to become "settled" is considerably reduced;
b. the configuration produces laminar flow conditions which enhance the settling rate and overall efficiency.

Tilted plates can also be used to uprate existing settling tanks.

Flotation may be used as an alternative to settlement. This process, which is gaining in popularity, consists in the carrying of metal hydroxides and other particles in suspension to the surface of the liquid in the flotation vessel by increasing particle buoyancy using the gas bubbles which adhere to the particles. The scum containing the gas bubbles and separated solids is skimmed off. Variations in the process lie mainly in the method of producing the carrier gas bubbles. This may be done by injecting a super-saturated solution of air in water under pressure into the tank - dissolved air flotation, or by injecting air through a diffuser - dispersed air flotation, or by the electrolysis of water to yield fine bubbles of hydrogen and oxygen - electrolytic flotation. The gas bubbles produced in these processes are extremely small, normally in the range 70-150 μm.

The use of direct filtration appears to be a very attractive process for the phase separation but unfortunately is seldom appropriate, mainly because of the tendency for the filter media to blind rapidly. This tendency is largely due to the gelatinous nature of the metal hydroxide precipitates. Occasionally where a more granular precipitate is obtained, direct filtration can be satisfactory and a high quality effluent can be obtained.

Whilst filtration has only limited application as the main means of solids removed, it is frequently used to polish the effluent from a settlement or flotation system to produce a higher quality effluent.

Where the metal is substantially in solution there are various techniques for separation or concentration of the metal so that a high quality treated effluent may be obtained.

a. Ion exhange

Ion exchange resins are in general of two types, insoluble organic acids used for cation exchange and insoluble organic bases used for anion exchange. Cation exchangers may be either sulphonic or carboxylic acids while anion exchangers may be either quaternary or tertiary amines.

An ion exchange system consists of a pair of columns or pressure vessels, one containing an anion exchange resin, the other the cation exchange resin. The effluent is continuously pumped through the two columns in series to yield the treated effluent. Where an exceptionally high quality effluent is required a third 'mixed-bed' exchange column can be placed after the cation exchange column.

After a period of running the treated effluent quality deteriorates due to exhaustion of the resin capacity and the resins must be regenerated. It is usual to have two trains of ion exchange columns so that whilst one is being regenerated the other can take the flow.

It is clear that the regeneration liquors will contain all the metals and associated anions which the raw effluent originally contained but they will now be in a concentrated form. These liquors must be treated by the precipitation/ solids separation method but the relatively small volume frequently means that the treatment can be done on a batch basis and absolute control over the final effluent quality can be exercised.

Ion exchange systems are particularly suitable for the treatment of metal containing wastes where relatively low concentrations of metal have to be removed. This enables long periods between regeneration of the resins and the production of a good quality water which is usually suitable for re-use. Rinse waters from electroplating operations are frequently treated by this process.

In practice, in addition to metals, the waste may contain other contaminants which it may be necessary to remove to protect the ion exchange resins. Organic contaminants are particularly damaging and should be removed by passage of the effluent through an activated carbon column preceding the ion-exchange system.

b. Evaporation

Evaporation is one of the most common methods used in industry for the concentration of aqueous solutions. However, use of this process as a means for effluent treatment is rare and occurs only under special circumstances where the effluent contains a high concentration of a valuable material. The only application of note here is on the concentration of static rinses (drag out) from electroplating operations, especially chromium plating. In this application the rinse liquor is evaporated to a metal concentration which makes the concentrate suitable for direct reuse in the plating bath.

c. Reverse Osmosis

On a practical scale this process is still in its infancy and whilst several plants for the treatment of brackish water to yield potable water have been successfully installed, there are certain drawbacks which have limited its development for the treatment of industrial effluents. In particular the process requires high pressures (up to 100 atmospheres) and is thus fairly costly in terms of energy, and the delicacy of the membranes is restrictive with regard to solids content and pH of the material being treated. However, the process has been used on effluents from electroplating in the electronic components industry, and through the continuous development of the process and improved mechanical strength of membranes its range of applications will almost certainly increase.

d. Solvent Extraction

The separation of the components of a liquid mixture by treatment with a solvent in which one or more of the desired components is preferentially soluble is known as solvent extraction. It is a process which is widely used in the chemical and petrochemical industries but which has only recently been adopted for the recovery of metals from aqueous solutions. At present it finds only limited application in this sphere largely because of the cost of solvent loss.

e. Electrodialysis

The cost of this process is very dependent on dissolved solids concentration. It does not find common use in effluent treatment but may be appropriate in certain circumstances where the concentrate is of value. A recent development is a rotary electrode.

The foregoing discussion of the processes available for the treatment of metal-bearing effluents has been reviewed around the removal of metal salts rather than elemental metal. The metals themselves do arise in effluents from metal processing industries and metal fabrication operations. Where the metal is in large particles such as swarf, simple screening is usually satisfactory.

Smelting operations produce a very fine dust called 'fume' in the furnace off-gases. Where wet gas scrubbing of these gases is carried out, the resultant liquor which contains the fume must be treated in some way. Settlement is usually satisfactory although a polyelectrolyte may be necessary to achieve a satisfactory settling rate.

The conventional precipitation method described above yields a sludge containing 95-99% water. Even with extremely good consolidation in a deep tank, the sludge will still contain about 90% water. To reduce the bulk for disposal, it is usual to dewater by some mechanical means.

The items of equipment commonly used for this purpose are filter presses, rotary vacuum filters and centrifuges each of which has their own particular advantages and disadvantages. Both the rotary vacuum filter and centrifuge are continuous processes and their effectiveness is enhanced by the use of a coagulant to condition the sludge. The filter press is a batch process and whilst a coagulant may be added to aid filtration, a handleable cake containing up to 30% dry solids may be obtained on the neat sludge. In general sludge produced from a lime precipitation is more granular and has better dewatering characteristics than sludge produced using sodium hydroxide as the precipitating agent.

Disposal of metal bearing sludges is, without a doubt, a major problem in some areas if satisfactory tips are not available. Where toxic metals are involved the sludge must be disposed of on a tip which is sealed to prevent the pollution of groundwater or it may be incinerated and the residual ash sealed prior to dumping.

If the economics of metal recovery/sludge disposal can demonstrate that separation of wastes streams to recover specific metals is attractive, then this can ease the sludge disposal system. This can be the case in certain instances particularly with the integrated plating system and with specific ion exchange systems, e.g. chrome recovery. It seems likely that the rising costs for sludge disposal by contractors will probably tip the balance in favour of metal recovery.

BIBLIOGRAPHY

For those people unfamiliar with waste treatment the following references will provide more information on the principles and more detail on industrial waste treatment.

1. METCALF & EDDY (1979) Wastewater Engineering, 2nd Edition, McGraw Hill.
2. SAWYER, C.N. & McCARTHY, P.L. (1978) Chemistry for Environmental Engineering, 3rd Edition, McGraw Hill.
3. ECKENFELDER, W.W. & O'CONNOR, D.J. (1961) Biological Waste Treatment. Pergamon Press.
4. GURHAM, C.F. (1965) Industrial Wastewater Control, Academic Press.
5. TEARLE, K. editor (1973) Industrial Pollution Control. Business Books Ltd.
6. Symposium on the Treatment of Wastewaters from Chemical Industries, 22-27 September 1977. Department of Civil Engineering, University of Newcastle upon Tyne.
7. Symposium on Toxic Materials in Industrial Effluents by Society for Chemistry and Industry, Newcastle, October, 1980. Published in Chemistry and Industry, Number 8, April 1981, pages 253-296.
8. World Health Organization - Guidelines for the Control of Industrial Wastes. Notably -

   No. 5 Pulp and Paper Manufacture by J.C. Lamb
   No. 6 Metal Finishing Wastes by J.C. Lamb
   No. 9 Cotton Textile Wastes by T.R. Bashkaran

5
Sewage and Sewage Sludge Treatment

By J. N. Lester
DEPARTMENT OF CIVIL ENGINEERING, IMPERIAL COLLEGE, LONDON SW7 2BU, U.K.

1. INTRODUCTION

It is estimated that the volume of water used daily in England and Wales (exclusive of water abstracted for cooling purposes) amounts to 5,000 mil gal (23 mil m^3) or approximately 95 gal (430 l) per capita per day. Domestic use accounts for nearly 1,800 mil gal (8 mil m^3) of this average daily total. Nearly all the water used domestically and approximately 1,500 mil gal (6.8 mil m^3) of the water used by industry each day is discharged to the sewers, yielding a total sewage flow of 3,100 mil gal (14.1 mil m^3) or about 60 gal (275 l) per capita per day.

The sewage from approximately 44 million people in England and Wales is treated by conventional waste water treatment processes, that from about a further 6 million people is discharged without treatment to the sea and some 1 to 2 million people are not connected to the sewerage system. To achieve this degree of waste water treatment requires some 5,000 sewage treatment works serving populations in excess of 10,000; these are distributed throughout the ten Water Authorities in England and Wales. The sewerage systems which carry the sewage to the site of treatment, or point of discharge, are of two types. Foul sewers carry only domestic and industrial effluent. In areas serviced in this way there are entirely separate systems for the collection of stormwater which is discharged directly to natural watercourses. However, in older towns and cities considerable use has been made of combined foul and stormwater systems. The use of combined sewage systems leads to very significant changes in the flow of sewage during storms. However, even in foul sewers significant changes in the flow occur due to variations in the pattern of domestic and industrial water usage which is essentially diurnal, and at its greatest during the day. Infiltration will also influence the flow in the sewerage system. Although a properly laid sewer is watertight when constructed, ground movement and ageing may allow water to enter the sewer if it is below the water table. The combined total of average daily flows to a sewage treatment works is called the dry weather flow (DWF). The DWF is an important value in the design and operation of the sewage treatment works and other flows are expressed in terms of it. DWF is defined as the daily rate of flow of sewage (including both domestic and trade waste), together with infiltration, if any, in a sewer in dry weather. This may be measured after a period of 7 consecutive days during which the rainfall has not exceeded 0.25 mm.

The DWF may be calculated from the following formula.

$$DWF = PQ + I + E$$

where, P = population served
Q = average domestic water consumption (ld^{-1})
I = rate of infiltration (ld^{-1})
E = volume (in litres) of industrial effluent discharged to sewers in 24 hours

1.1 Objectives of sewage treatment.

Water pollution in the United Kingdom was already a serious problem by 1850. It is probable that the early endeavours to control water pollution were considerably stimulated by the state of the lower reaches of the River Thames which at the point where it passed the Houses of Parliament was grossly polluted. An early solution to these problems was sought through the construction of interceptor sewers. These collected all the sewage draining to the River Thames and carried it several miles down the river before discharging it to the estuary on the ebb tide. From there it moved towards the sea and in so doing received greater dilution. Despite these measures and the passing of the first Act of Parliament to control water pollution in 1876 the situation continued to deteriorate. The requirement for, and the objectives of, sewage treatment were first outlined by the Royal Commission on Sewage Disposal (1898-1915). The objectives of sewage treatment have developed significantly since this report; however, the standards described then are still applicable in many areas and this report provided the framework around which the United Kingdom waste water industry has developed.

Originally the objective of sewage treatment was to avoid pestilence and nuisance (disease and odour) and to protect the sources of potable supply.

During sewage treatment disease causing organisms may be destroyed or concentrated in the sludges produced; similarly offensive materials may be concentrated in the sludges or biodegraded. As a consequence the quantities of these agents present in the sewage effluent is much less than in the untreated sewage and their dilution in the receiving water far greater. The benefits of sewage treatment are not limited to greater dilution however, since each receiving water has a certain capacity for "self purification". Providing sewage treatment reduces the burden of polluting material to a value less than this capacity then the ecosystem of the receiving water will complete the treatment of the residual materials present in the sewage effluent. Thus sewage treatment in conjunction with the selection of appropriate points for sewage effluent discharge has resulted in the elimination of waterborne disease in the UK and many other advanced countries. However, as the population has expanded and become urbanised with a concommitant development of water-consuming industries an additional requirement has been placed upon sewage treatment.

It is now the objective of sewage treatment in many parts of the UK to produce a sewage effluent which after varying degrees of dilution and self purification is suitable for abstraction for treatment to produce a potable supply. This indirect reuse affects some 30% of all water supplies in the UK.

1.2 The importance of water re-use.

That the United Kingdon practises indirect re-use to a greater extent than most other countries may appear surprising given the annual rainfall. Indeed that re-use should be important in global terms given the abundance of water on the earth's surface may also be considered improbable in all but the most arid regions. However, two important factors readily explain this situation: a vast amount of the available water is too saline to be used as a potable supply (the salinity is too costly to remove in all but the most extreme cases) and secondly the non-uniform distribution of population and the available water supply. The available water supply is determined by the rainfall, the ability of the environment to store water (essentially the size of lakes and rivers, which are small in the United Kingdom) and their location, i.e. Wales has an abundance of suitable water supplies, but limited population, whilst South East England has a large population with limited water resources.

It has been estimated that of the water falling on the United Kingdom 50% is not available for use as a result of run-off to the sea. Of the remainder approximately 17% is utilised. Current predictions suggest that by 2000 AD the amount of water used will have doubled. Thus the potential reserves are very limited. However, because demand and supply are not geographically proximate re-use is already essential. As a consequence the traditional concept of water supply employing single-purpose reservoirs impounding unused river water has been abandoned in favour of multi-purpose schemes designed to permit repeated use of the water before it reaches the sea. In these schemes sewage treatment plays a vital role in addition to being an integral part of the hydrological cycle.

1.3 Criteria for sewage treatment.

Sewage is a complex mixture of suspended and dissolved materials; both categories constitute organic pollution. The strength of sewage and the quality of sewage effluent are described in terms of their Suspended Solids (SS) and Biochemical Oxygen Demand (BOD): these two measures were either proposed or devised by the Royal Commission (1898-1915).

The SS are determined by weighing after the filtration of a known volume of sample through a standard glassfibre filter paper; the results are expressed in mgl^{-1}.

Dissolved pollutants are determined by the BOD they exert when incubated for 5 days at $20^{o}C$. Samples require appropriate dilution with oxygen-saturated water and suitable replication. The oxygen consumed is determined and the results again expressed in mgl^{-1}.

The two standards for sewage effluent quality proposed by the Royal Commission were for no more than 30 mgl^{-1} of suspended solids and 20 mgl^{-1} for BOD, the so called 30:20 standard. The Royal Commission envisaged that the effluent of this standard would be diluted 8:1 with clean river water having a BOD of 2 mgl^{-1} or less. This standard was considered to be the normal minimum requirement and was not enforced by statute because the character and uses of rivers varied so greatly. It was intended that standards would be introduced locally as required. For example, rivers to be used for abstraction of potable supplies would require a higher standard such as the 10:10 standard imposed by the Thames Conservancy. Whilst other countries which are members of the European Economic Community have adopted "uniform emission standards", that is the same quality of effluent regardless of the state or use of the river, the United Kingdom has continued with its pragmatic approach whereby effluent standards are set depending on the "water quality objectives" of the river, which in turn is determined by its function or use. In the 1970's with the reorganisation the water industry's reliance solely on the 30:20 standard has been abandoned, although this standard is probably still the most commonly applied. Sewage treatment now attempts to produce consistently an effluent with a quality superior to its "Legal Consent" and attempts to achieve an "Operating Target" frequently half the Legal Consent. In addition considerable importance has been placed upon the concentration of ammonia in the effluent. In the case of a works attempting to nitrify the effluent (see section 2.3.3) the ammonia concentration is frequently limiting. Typical Legal Consent and Operating Targets are outlined in Table 1.

Table 1.

Legal Consent and Operating Target values for a conventional two stage sewage treatment works.

Parameter	Legal Consent value (mgl^{-1})	Operating Target value (mgl^{-1})
SS	50	30
BOD	35	20
ammonia	25	12

It is evident that the Operating Targets included in Table 1 are the same as the Royal Commission 30:20 standard.

1.4 Composition of sewage.

Domestic sewage contains approximately 1000 mgl^{-1} of impurities of which about two thirds are organic. Thus sewage is 99.9% water and 0.1% total solids upon evaporation. When present in sewage approximately 50% of this material is dissolved and 50% suspended (see Figure 1). The main components are: nitrogenous compounds - proteins and urea; carbohydrates - sugars, starches and cellulose; fats - soap, cooking oil and greases. Inorganic components include chloride, metallic salts and road grit where combined sewerage is used.

Figure 1.

Composition of a typical raw sewage.

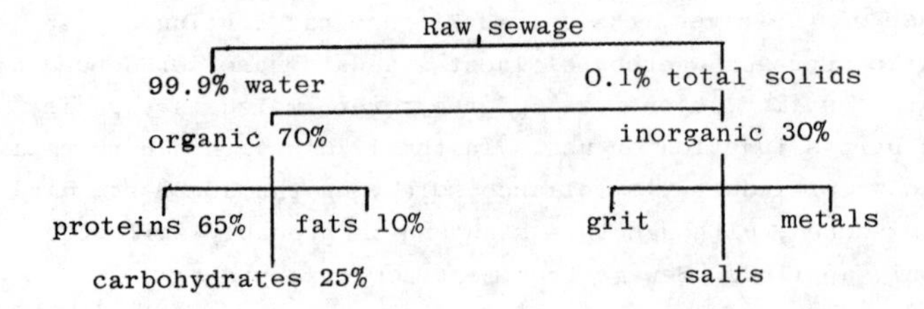

Thus sewage is a dilute, heterogenous medium which tends to be rich in nitrogen.

2. SEWAGE TREATMENT PROCESSES

Conventional sewage treatment is a three stage process including preliminary treatment, primary sedimentation and secondary (biological) treatment; these are presented schematically in Figure 2. In addition some form of sludge treatment facility is frequently employed, typically anaerobic digestion (see section 3.1).

Figure 2.

Flow diagram of a conventional Sewage Treatment works.

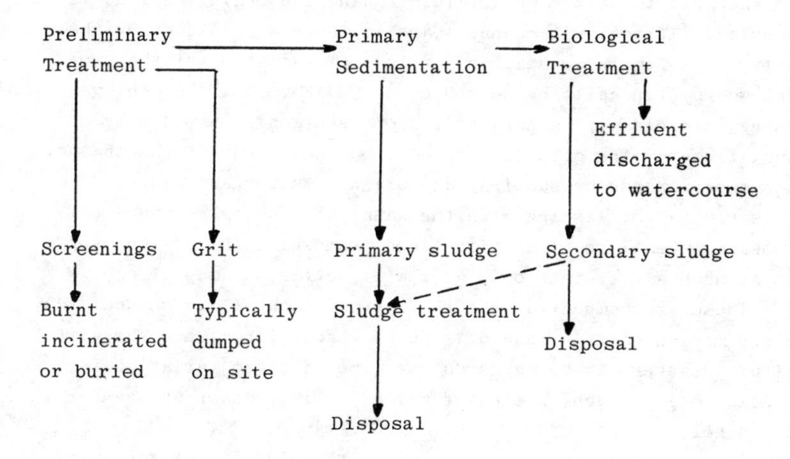

2.1 Preliminary Treatment

These treatment processes are intended to remove the larger floating and suspended materials. They do not make a significant contribution to reducing the polluting load, but render the sewage more amenable to treatment by removing large objects which could form blockages or damage equipment.

Floating or very large suspended objects are frequently removed by bar screens. These consist of parallel rods with spaces between them which vary from 40 to 80 mm, through which the influent raw sewage must pass. Material which accumulates on the screen may be removed manually with a rake at small works, but on larger works some form of automatic raking would be used. The material removed from the screens contains a significant amount of putrescible organic matter which is objectionable in nature and may pose a disposal problem. Typically the material is buried or incinerated and less frequently burnt.

If screens have been used to remove the largest suspended and virtually all the floating objects, then it only remains to remove the small stones and grit, which may otherwise damage pumps and valves, to complete the preliminary treatment. This is most frequently achieved by the use of constant velocity grit channels. The channels utilise differential settlement to remove only the heavier grit particles whilst leaving the lighter organic matter in suspension. A velocity of 0.3 ms^{-1} is sufficient to allow the grit to settle whilst maintaining the organic solids in suspension. If the grit channels are to function efficiently the velocity must remain constant regardless of variation of the flow to the works (typically between 0.4 and 9 DWF). This is achieved by using channels with a parabolic cross section controlled by venturi flumes. The grit is removed from the bottom of the channel by a bucket scraper or suction, and organic matter adhering to the grit is removed by washing with the wash water being returned to the sewage. Small sedimentation tanks from which the sewage overflows at such a rate that only grit will settle out may also be used. These are compact and by the introduction of air on one side a rotary motion can be induced in the sewage which washes the grit *in situ*. However, these tanks do not cope with the variation in hydraulic load in such an elegant and effective manner as the grit channel.

To avoid the problems associated with the disposal of screening comminutors are frequently employed in place of screens. Unlike screens which precede grit removal the comminutors are placed downstream of the grit removal process. The comminutors shred the large solids in the flow without removing them. As a result they are reduced to a suitable size for removal during sedimentation. Comminutors consist of a slotted drum through which the sewage must pass. The drum slowly rotates carrying material which is too large to pass through the drum towards a cutting bar upon which it is shredded before it passes through the drum.

The total flow reaching the sewage treatment works is subjected to both these preliminary treatment processes. However, the works is only able to give full treatment up to a maximum flow of 3 DWF.

When the flow to the works exceeds this value the excess flows over a weir to the storm tanks which are normally empty. If the storm is short no discharge occurs and the contents of the tanks are pumped back into the works when the flow falls below 3 DWF. If the storm is prolonged then these tanks will begin to discharge to a nearby watercourse, inevitably causing some pollution. However, this excess flow has been subjected to sedimentation which removes some of the polluting material.

2.2 Primary Sedimentation

The raw sewage (containing approximately 400 mgl^{-1} SS and 300 mgl^{-1} BOD), at a flowrate of 3 DWF or less and with increased homogeneity as a result of the preliminary treatment processes, enters the first stâge of treatment which reduces its pollutant load, primary sedimentation or mechanical treatment. Circular (radial flow) or rectangular (horizontal flow) tanks equipped with mechanical sludge scraping devices are normally used (see Figure 3). However, on small works hopper bottom tanks (vertical flow) are preferred; although more expensive to construct these costs are most than offset by savings made as a result of eliminating the requirement for scrapers (see Figure 3).

Removal of particles during sedimentation is controlled by the settling characteristics of the particles (their density, size and ability to flocculate), the retention time in the tank (h), the surface loading ($m^3m^{-2}d^{-1}$) and to a very limited degree the weir overflow rate ($m^3m^{-1}d^{-1}$). Retention times are generally between 2 and 6 h. However, the most important design criterion is the surface loading; typical values would be in the range 30 to 45 $m^3m^{-2}d$. The surface loading rate is obtained by dividing the volume of sewage entering the tank each day (m^3d^{-1}) by the surface area of the tank (m^2). The retention time may be fixed independently of the surface loading by selection of the tank depth, typically 2 to 4m, which increases the volume without influencing the surface area. Because they strongly influence the value for surface loading selected, the nature of the particles in the sewage is one of the most important factors in determining the design and efficiency of the sedimentation tank. Of the three factors mentioned before, flocculation is perhaps the most significant.

Figure 3

Types of sedimentation tank

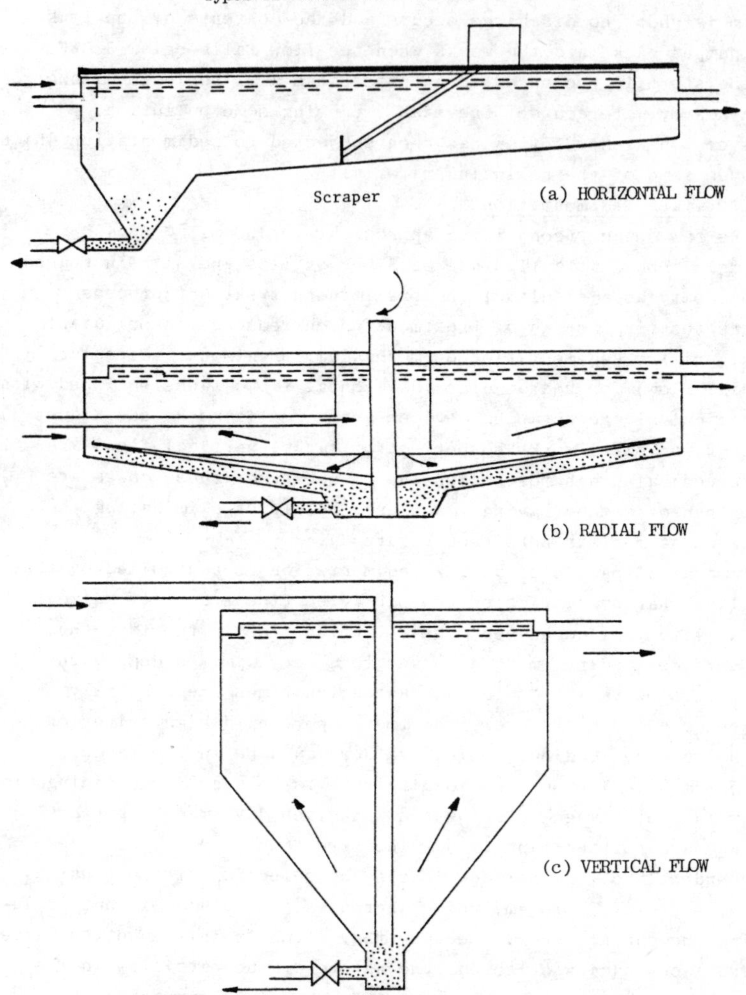

Four different types of settling can occur:

Class 1 Settling: settlement of discrete particles in accordance with theory (Stokes' Law).

Class 2 Settling: settlement of flocculant particles exhibiting increased velocity during the process.

Zone Settling (Hindered settlement): at certain concentrations of flocculant particles, the particles are close enough together for the interparticulate forces to hold the particles fixed relative to one another so that the suspension settles as a unit.

Compressive Settling: at high solids concentrations the particles are in contact and the weight of the particles is in part supported by the lower layer of solids.

During primary sedimentation settlement is of the Class 1 or 2 types. However, in secondary sedimentation (see section 2.4) zone or hindered settlement may occur. Compressive settlement only occurs in special sludge thickening tanks.

Primary sedimentation removes approximately 55% of the suspended solids and because some of these solids are biodegradable the BOD is typically reduced by 35% The floating scum is also removed and combined with the sludge. As a result the effluent from the primary as a SS of approximately 150 mgl^{-1} and a BOD of approximately 200 mgl^{-1}. This may be acceptable for discharge to the sea or some estuaries without further treatment. The solids are concentrated into the primary sludge which is typically removed once a day under the influence of hydrostatic pressure.

2.3 Secondary (Biological) Treatment

There are two principal types of biological sewage treatment,

i. The percolating filter (also referred to as a trickling or biological filter).

ii. Activated sludge treatment.

Both types of treatment utilise two vessels, a reactor containing the micro-organisms which oxidise the BOD, and a secondary sedimentation tank, which resembles the circular radial flow primary sedimentation tank, in which the micro-organisms are separated from the final effluent.

The early development of biological sewage treatment is not well documented. However, it is established that the percolating filter was developed to overcome the problems associated with the treatment of sewage by land at "sewage farms", where large areas of land were required for each unit volume of sewage treated. It was discovered that approximately 10 times the volume of sewage could be treated in a given area per unit time by passing the sewage through a granular medium supported on underdrains designed to allow the access of air to the microbial film coating the granular bed.

The origins of the percolating filter are present in land treatment and its development was an example of evolution. The second and probably predominant form of biological sewage treatment, the activated sludge process, arose spontaneously and represents an entirely original approach. This process involves the aeration of freely suspended flocculant bacteria, "the activated sludge floc" in conjunction with the settled sewage which together constitute the "mixed liquor". Activated sludge treatment continues the trend established by the change from land treatment to the percolating filter in that at the expense of higher operating costs it is possible to treat very much larger volumes of sewage in a smaller area.

The activated sludge process is probably the earliest example of a continuous bacterial (microbial) culture deliberately employed by man, and certainly the largest used to date. Development of the activated sludge process was announced by its originators Fowler, Ardern and Lockett in 1913, based upon their research at the Davyhulme sewage treatment works, Manchester. These scientists very generously did not patent the process to facilitate its rapid and widescale application.

Development of these two forms of biological sewage treatment has been largely empirical and undertaken without the benefit of information about the fundamental principles of continuous bacterial growth, which began to be developed from the late 1940's when Monod published his work on continuous bacterial growth, although the relevance was not perceived until approximately 10 years later. This lack of microbiological knowledge is highlighted by the fact that the role of micro-organisms in the activated sludge process was not fully accepted until after 1931; prior to this it was accepted by several workers that coagulation of the sewage colloids was the principal mechanism in the activated sludge process, although in the USA the role of bacteria in percolating filters was first recognised in 1889.

Figure 4
A percolating filter

(a) SECTION

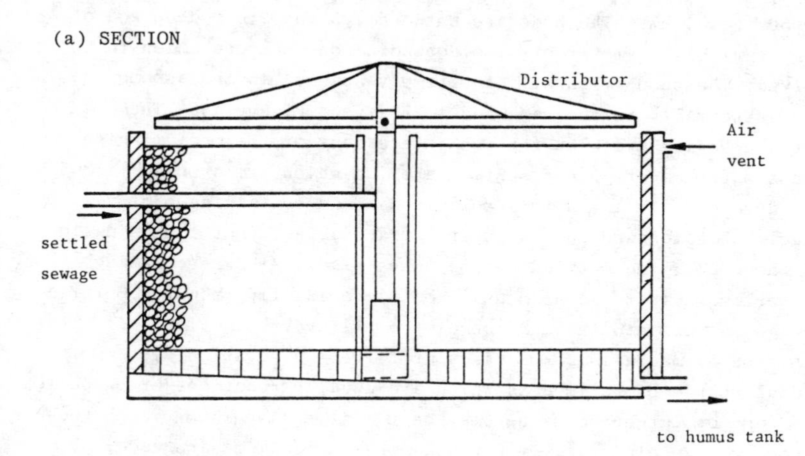

(b) PLAN

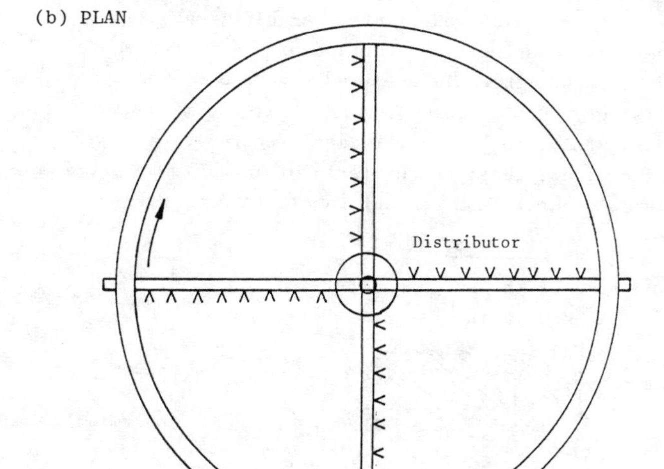

2.3.1 Percolating Filter

These units consist of circular or rectangular beds of broken rock, gravel, clinker or slag with a typical size in the range of 50 to 100 mm. The beds are between 1.5 and 2.0 m deep and of very variable diameter or size depending on the population to be served. The proportion of voids (empty spaces) in the assembled bed is normally in the range 45 to 55% (see Figure 4). The settled sewage trickles through interstices of the medium which constitutes a very large surface area on which a microbial film can develop. It is in this gelatinous film containing bacteria, fungi, protozoa and on the upper surface algae, that the oxidation of the BOD in the settled sewage takes place. The percolating filter is in fact a continous mixed microbial film reactor.Settled sewage is fed onto the surface of the filter by some form of distributor mechanism. On circular filters a rotation system of radial sparge pipes is used which are usually reaction-jet propelled although on larger beds they may be electrically driven. With rectangular beds electrically powered rope-hauled arms are used.

The micro-organisms which constitute the gelatinous film appear to be organised, at least near the surface of the filter where algae are present, into three layers (see Figure 5). The upper fungal layer is very thin (0.33 mm), beneath it the main algal layer is approximately 1.2 mm and both are anchored by a basal layer containing algae, fungi and bacteria of approximately 0.5 mm. However, algae do occur to some extent in all three layers.

Figure 5.

Cross-section of the surface layers of a percolating filter.

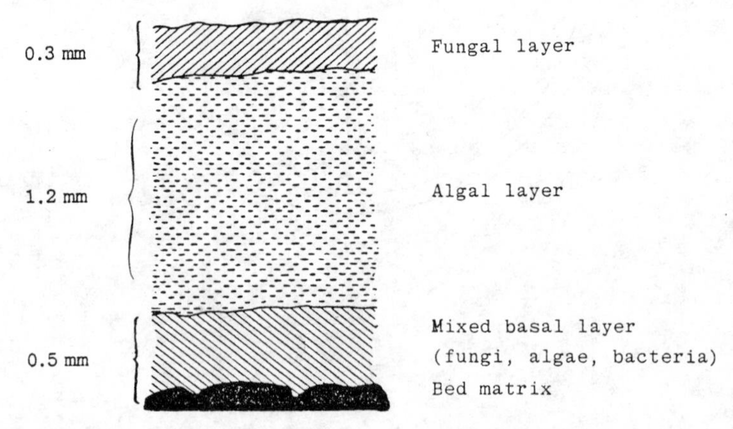

Beneath the surface where sunlight is excluded and as a consequence the algae are absent, this structure is significantly modified, probably into a form of organisation with only two layers. It has been calculated that photosynthesis by algae could provide only 5% or less of the oxygen requirements of the micro-organisms in the filter. Furthermore, photosynthesis would only be an intermittent source of oxygen since it would not occur in the dark and algae are often present only in the summer months. Carbon dioxide generated by other organisms in the filter might increase the rate of photosynthesis. It has been proposed that algae derive nitrogen and minerals from the sewage and that some may be facultative heterotrophs. The nitrogen fixing "blue-green" algae are frequently present in filters.

Whilst fungi are efficient in the oxidation of the BOD present in the settled sewage they are not desirable as dominant members of the microbial community. Per unit of BOD consummed, they generate more biomass than bacteria, thus increasing the sludge disposal problem. Moreover, an accumulation of predominantly fungal film quickly causes blockages of the interstices of the filter bed material, impeding both drainage and aeration. The latter may result in a reduction in the efficiency of treatment which is dependent upon the metabolic activity of aerobic micro-organisms.

Protozoa and certain metazoans (macrofauna) play an important role in the successful performance of the biological filter, although the precise nature of this role is dependent on the extrapolation of observations made in the activated sludge process, which is more amenable to study. However, the similarity in the distribution of organisms within the two processes suggests strongly that their roles are the same in both. The protozoa in particular remove free-swimming bacteria thus preventing turbid effluents, since freely suspended bactera are not settleable. Certain metazoans may also ingest free-swimming bacteria, but their most important function is to assist in breaking the microbial film which would otherwise block the filter. This film is "sloughed off" with the treated settled sewage. Protozoa (principally ciliates and flagellates) tend to dominate in the upper layers of the filter, whilst the macro-fauna (nematodes, rotifers, annelids and insect larvae) dominate the lower layers.

If film is not removed satisfactorily frequently as the result of excessive fungal growth the condition known as "ponding" develops. In this condition the surface of the filter is covered in settled sewage, air flow ceases, treatment stops and the bed becomes anaerobic. Ponding may also be caused by the growth of a sheet or felt of large filamentous algae, principally Phormidiam sp., on the face of the filter.

To minimise film production recirculation of treated effluent is often employed. This reduces film growth by dilution of the settled sewage, improves the flushing action for the removal of loose film and promotes more uniform distribution of the film with depth.

Treated sewage is subject to secondary sedimentation which is similar to primary sedimentation as a result of which the suspended sloughed off film is consolidated into humus sludge and the final effluent discharged to the receiving water.

2.3.2. Activated sludge

In the activated sludge process the majority of biological solids removed in the secondary sedimentation tank are recycled (returned sludge) to the aerator. The feedback of most of the cell yield from the sedimentation tank encourages rapid adsorption of the pollutants in the incoming settled sewage and also serves to stabilise operation over the wide range of dilution rates and substrate concentrations imposed by the diurnal and other fluctuations in the flow and strength of the sewage. Stability is also provided by the continuous inoculation of the reactor with micro-organisms in the sewage and airflows, which are ultimately derived from human and animal excreta, soil run-off, water and dust. The reactor of the activated sludge plant is usually in the form of long deep channels. Before entering these channels the returned sludge and settled sewage are mixed thereby forming the "mixed liquor". The retention time of the mixed liquor in the aerator is typically 3 to 6 hours; during this period it moves down the length of the channel before passing over a weir, prior to secondary sedimentation. The sludge which is not returned to the aerator unit is known as surplus activated sludge and has to be disposed of. In practice the conditions in the aeration unit diverge from the completely mixed conditions commonly used for industrial fermentations and it may be best described as a continuous mixed microbial deep reactor with feed-back.

The design of the concrete tanks which form the reactor is strongly influenced by the type of aeration to be employed. Two types are available, compressed (diffused air) (see Figure 6) and mechanical (surface aeration) (see Figure 7). In the diffused air system much of the air supplied is required to create turbulence, to avoid sedimentation of the bacteria responsible for oxidation. Surface aeration systems introduce the turbulence mechanically and only provide sufficient air for bacterial oxidation. Both types of system aim to maintain a dissolved oxygen concentration of between 1 and 2 mgl^{-1}.

In the diffused air system the air is released through a porous sinter at the base of the tank and this system is characterised by long undivided channels which may be quite narrow (see Figure 6). Mechanical aeration utilises rotating paddles to agitate the surface thereby incorporating air and creating a rotating current which maintains the bacterial flocs in suspension. Each paddle is located in its own cell which has a hopper shaped bottom; this gives the plant the appearence of a square lattice (see Figure 7). However beneath the face of the mixed liquor all the cells are connected forming a channel. In both systems the channels are 2 to 3 m deep and 40 to 100 m long.

The success of the activated sludge process is dependent on the ability of the micro-organisms to form aggregates (flocs) which are able to settle. It is generally accepted that flocculation can be explained by colloidal phenomena and that bacterial extracellular polymers play an important role, but the precise mechanism is not known. The significance of flocculation to the success of the process is not the only characteristic to distinguish it from other industrial continuous cultures. There are four additional and very significant differences: it utilises a heterogenous microbial population, growing in a very dilute multisubstrate medium, many of the bacterial cells are not viable and finally the objectives of the process, which are the complete mineralisation of the substrates (principally carbon dioxide water, ammonia and/or nitrate) with minimal production of both biomass and metabolites, are also unique.

The heterogenous population present in activated sludge includes bacteria, protozoa, rotifers, nematodes and fungi. The bacteria alone are responsible for the removal of the dissolved organic material, whilst the protozoa and rotifers "graze", removing any

Figure 6

Diffused aeration activated sludge plant

(a) SCHEMATIC

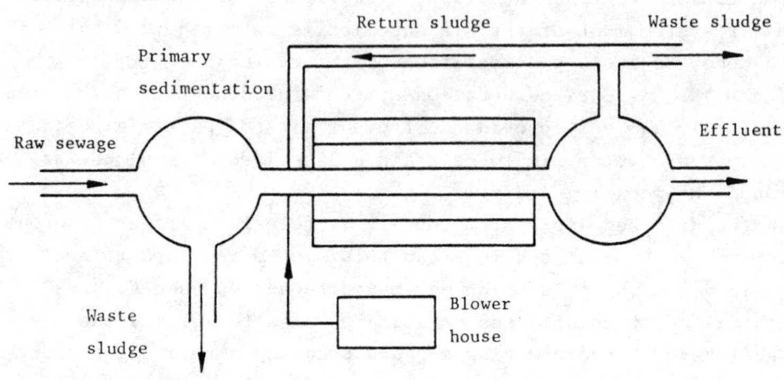

(b) SECTION

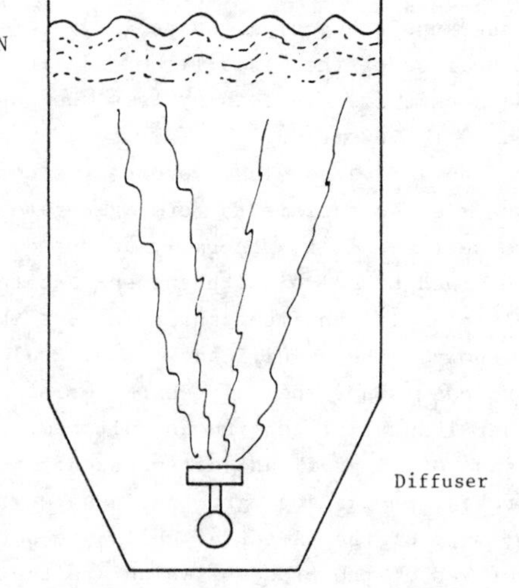

Figure 7

Mechanically aerated activated sludge plant

(a) SCHEMATIC

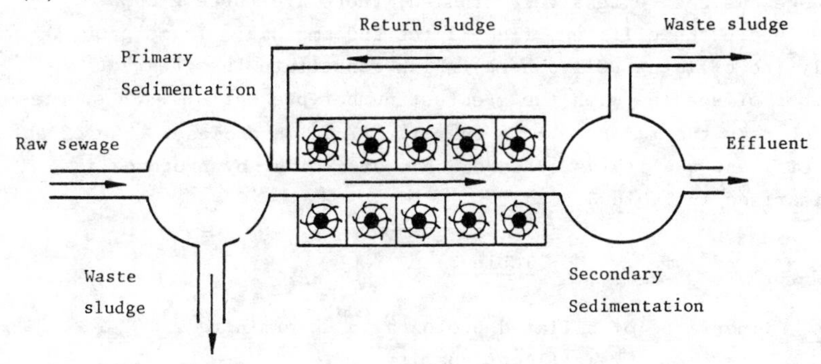

(b) SECTION

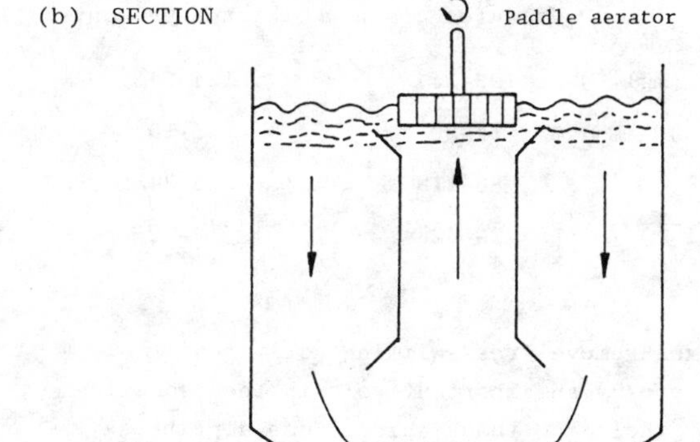

"free-swimming" and hence non-settleable bacteria, the protozoans and rotifers being large enough to settle during secondary sedimentation. The role of protozoa in activated sludge has been extensively studied; there are three groups involved: the ciliates, flagellates and amoebae. It is probably the ciliates (Ciliophora) which constitute the greatest number of species with the greatest number present in each species which play the major role in the clarification process. The effect on effluent quality as a consequence of grazing by protozoa is summarised in Table 3.

Table 3.

Importance of ciliated protozoa in determining effluent quality

Effluent Property	Ciliates absent	Ciliates present
Chemical Oxygen Demand (mgl^{-1})	198-254	124-142
Organic Nitrogen (mgl^{-1})	14-20	7-10
Suspended solids (mgl^{-1})	86-118	26-34
Viable bacteria (10^7 ml^{-1})	29-42	9-12

Not only do the protozoa remove free-swimming activated sludge bacteria but they play an important role in the reduction of pathogenic bacteria, including those which cause diphtheria, cholera, typhus and streptococcal infections. In the absence of protozoa approximately 50% of these types of organisms are removed while in their presence removals rise to 95%. Nematodes have no significant role in the process, whilst the effects of the fungi are generally deleterious and contribute to or cause non-settleable sludge known as "bulking". Members of the following bacterial genera have been regularly isolated from activated sludge, *Pseudomonas*, *Acinetobacter*, *Comamonas*, *Lophomonas*, *Nitrosomonas*, *Zoogloea*, *Sphaerotilus*, *Azotobacter*, *Chromobacterium*, *Achromobacter*, *Flavobacterium*, *Alcaligenes*, *Micrococcus* and *Bacillus*. Attributing the appropriate importance to each genus is a problem which confounds bacteriologists.

Of the principal groups of substrates listed in Section 1.4, only one single substrate (cellulose) was included. Each of the groups include many substrates: for example the "sugars" identified in sewage include glucose, galactose, mannose, lactose, sucrose, maltose and arabinose, whilst the nitrogenous compounds include proteins, polypeptides, peptides, amino-acids, urea, creatine and amino-sugars. Since bacteria normally only utilise a single carbon substrate or at the most two, this diversity of substrates in part explains the numerous genera of bacteria isolated from activated sludge, because each substrate under most conditions will sustain one species of bacterium. Moreover as a consequence of the large number of substrates present in the settled sewage the concentration of individual substrates is far less than the 200 mgl^{-1} of BOD present, perhaps 20 to 40 mgl^{-1} for the most abundant and less than 10 mgl^{-1} for the uncommon ones. The concentration of each substrate is further reduced in the aeration tank by dilution with the returned activated sludge, which is typically mixed 1:1 with settled sewage resulting in a 50% reduction in substrate concentration.

The low substrate concentration means that the bacteria are in a starved condition. As a consequence many of them are "senescent" i.e. in that phase between death as expressed by the loss of viability, and breakdown of the osmotic regulatory system (the moribund state); thus the bacterium is a functioning biological entity incapable of multiplication. That bacteria could exist in this condition was established at an early stage in a series of inspired experiments by Wooldridge and Standfast who published their results in 1933. They determined the dissolved oxygen concentration and bacterial numbers (by viable counts) in a series of Biochemical Oxygen Demand bottles containing diluted raw sewage, on a daily basis. The viable count reached a maximum on the second day and thereafter fell rapidly. However the consumption of oxygen increased by equal amounts until the fourth day and fell to a negligible value on the fifth day. There was no obvious relationship between viability and oxygen consumption. They tested experimentally the hypothesis that non-viable bacteria were apparently capable of oxygen uptake by destroying the capacity for division without significantly diminishing enzyme activity. Treatment of *Pseudomonas fluorescens* with a 0.5% formaldehyde solution prevented division, but these bacteria exhibited vigorous oxygen uptake in both sewage and other media. Susequently they were able to determine the presence of active oxidase and dehydrogenase enzymes in these non-viable bacteria. The effects of low substrate concentration on the viability of the bacteria are compounded by their specific growth rate. It is intended that biological waste water treatment should result in the production of a final effluent containing negligible BOD. The biochemically oxidisable material in the effluent is composed of compounds originally present in the settled sewage, which have not been completely biodegraded, and bacterial products. Moreover, this is to be achieved with the minimal production of biomass. These twin objectives are concomittant with the utilisation of a bacterial population with a very low specific growth rate.

Unlike the percolating filter bacterial growth in the activated sludge process is amenable to the type of description used by bacteriologists for conventional continuous cultures. However, although it is amenable to this type of treatment it inevitably appears to be very different from all other continuous cultures. The dilution rates (rate of inflow of settled sewage/aeration tank volume) used are invariably low by the standards of industrial fermentations, typically 0.25 h^{-1}, i.e. one quarter of the aeration tank volume is displaced every hour; therefore the hydraulic retention time is four hours. In the conventional single pass reactor the dilution rate and the specific growth rate (time required for a doubling of the population) are identical. That is the state in which the rate of production of cells through growth equals the rate of the loss of cells through the overflow. In the activated sludge process because of the recycling of the biomass the specific growth rate is very much lower than the dilution rate, typically in the range 0.002 to 0.007 h^{-1}. Since under steady state conditions the bacteria are only able to grow at the same rate as they are lost from the system, recycling them dramatically lowers their specific growth rate and allows it to be controlled independently of the dilution rate. Under steady state conditions the specific growth rate is equivalent to the specific rate of sludge wastage (mass of suspended solids lost by sludge wastage and discharged in the effluent in unit time as a proportion of the total mass in the plant) which is the reciprocal of the "sludge age" or mean cell retention time which is typically 4 to 9 days. Thus whilst the retention of the aqueous phase in the system is only 4 h the retention of the bacterial cells or sludge age is several days. The sludge age (θc) is a value which describes a great deal about the type of activated sludge plant; its purpose, quality of effluent and bacteriological and biochemical states are all summarised by this term.

The activated sludge process may have up to four phases:

i) clarification, by flocculation of suspended and colloidal matter;

ii) oxidation of carbonaceous matter;

iii) oxidation of nitrogenous matter; (see section 2.3.3)

iv) auto-digestion of the activated sludge.

The occurrence of these four phases is directly dependent on increasing sludge age. These processes which operate at low sludge ages give rapid removal of BOD per unit time, but the effluent is of poor quality. Plants which have high sludge ages give good quality effluents but only a slow rate of removal. Low sludge ages result in actively growing bacteria and consequently high sludge production, whilst bacteria grown at high sludge ages behave conversely. Figure 8 illustrates the relationships between the growth curve of the bacterial culture and the type of activated sludge plants. By operating continuously the activated sludge process functions only over a small region of the batch; this region is determined by the specific sludge wastage rate. The region selected determines the type of plant and its performance. These are summarised below.

Figure 8.

Relationship between batch culture and type of activated sludge plant

carbon oxidation
autodigestion
Function
adsorption
nitrogen oxidation
lag phase — exponential phase — stationary phase — death phase — Bacteriological growth state
high rate activated sludge
Types of activated sludge plant
dispersed aeration
conventional activated sludge
extended aeration plant
substrate concentration →
endogenous respiration
no. of organisms →
Time

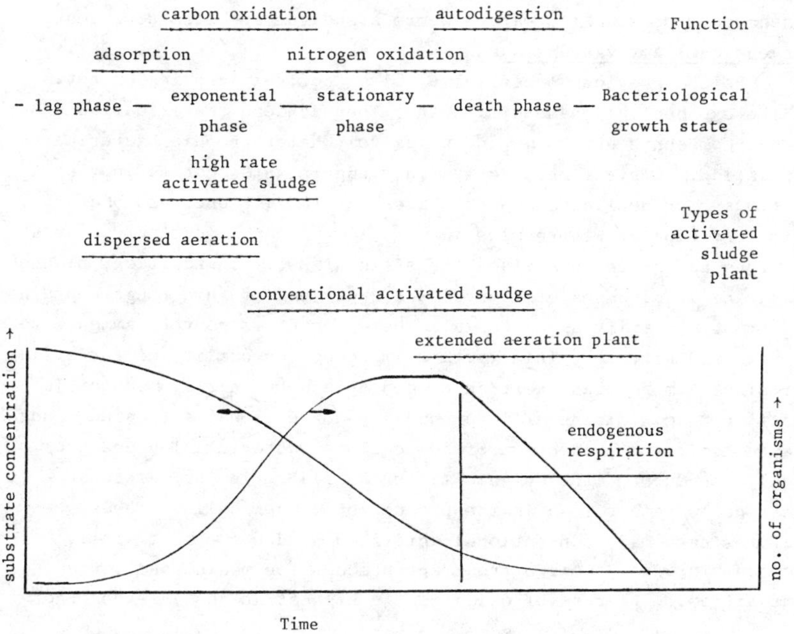

Dispersed aeration

This type of process is rarely used and is not applicable to the treatment of municipal sewage, but may be of use in the preliminary treatment of some industrial wastes. The bacteria are growing rapidly (exponential phase); thus the process has the ability to remove a large quantity of BOD per unit of biomass and as a consequence a small reactor may be used which is cheap to construct. However because of their high rate of growth the bacteria convert much of the BOD into biomass, causing a sludge disposal problem. Flocculation is limited so additional treatment is essential to remove solids. Furthermore, although BOD removal per unit biomass is high the effluent BOD is also high.

High-rate activated sludge

This shares many features of the previous process , but flocculation proceeds satisfactorily and secondary sedimentation will remove the solids effectively. The growth rate of the bacteria is still high, and only carboneous material will be oxidised. However some 60 to 70% of the influent BOD will be removed with a hydraulic retention time of approximately 2 h. This type of process is probably most frequently used for industrial wastes prior to discharge to the sewers, although it is also used for domestic sewage treatment perhaps most appropriately where effluents are to be discharged to estuarine waters where standards are less stringent.

Conventional activated sludge

The two previous processes utilise actively growing bacteria in the exponential phase of growth. They achieve the oxidation of carbon compounds utilising an exclusively heterotrophic bacterial population. Conventional activated sludge plants operate in the stationary or declining growth phases, utilising senescent bacteria. This very slow growth results in very low residual substrate concentrations and hence low values for effluent BOD. In addition, plants operating at sludge ages towards the upper end of this range contain autotrophic nitrifying bacteria. These organisms convert ammonia to nitrite and nitrate. This further improves the quality of the effluent since ammonia can exert an oxygen demand but nitrate cannot.In addition to maximising effluent quality conventional activated sludge plants limit the production of new cells. Bacteria which are growing slowly use much of the organic matter available in the maintenance of their cells rather than in the production of new cells. These features have made conventional activated sludge the most widely adopted biological sewage treatment process for medium and large communities. The rate of oxidation is highest at the inlet of the tank and it can be difficult to maintain aerobic conditions. Two solutions to this problem have been adopted. With tapered aeration rather than supplying air uniformly along the length of the tank the air is concentrated at the beginning of the tank and progressively reduced along its length. The volume of air supplied remains unchanged but it is distributed according to demand. Alternatively stepped loading may be utilised. This aims to make the requirement for air uniform by adding the settled sewage at intervals along the tank, thus distributing the demand.

Extended aeration

This process operates at very high sludge ages exclusively in the declining phase of growth. The retention time in the aeration tank is between 24 h and 24 days; as a consequence the available substrate concentration is low and the bacteria undertake endogenous respiration (see Figure 8), that is respiration after the consumption of all available extracellular substrate. The result of utilising endogenous materials is the breakdown of the sludge, sometimes referred to as auto-digestion. By this means sludge production is minimised and the small amount of material that must be disposed of is highly mineralised and inoffensive. This type of treatment has been extensively used for small communities, whilst capital costs of such plants are high operating and sludge disposal costs are very low.

Contact stabilisation

The contact stabilisation process is a variation of conventional activated sludge used for treating wastes with a high content of biodegradable colloidal and suspended matter. The process utilises the adsorptive properties of the sludge to remove the polluting material very rapidly (0.5 to 1 h) in a small aeration tank. The mixed liquor is then settled and passed into a second aeration tank and aerated for a further 5 to 6 h, during which period the adsorbed material is oxidised. After this the sludge with its adsorptive capacity restored is returned to the contact basin. Although this process requires two aeration tanks the two are very much smaller than the equivalent single tank since the mixed liquor suspended solids in the contact basin are typically 2000 mgl^{-1} and in the second tank (digestion unit) they are about 20,000 mgl^{-1}

2.3.3 Nitrification

The production of a final effluent with the minimum BOD value is dependent upon the complete nitrification of the effluent, which involves the conversion of the ammonia present to nitrate. This is a two stage process undertaken by autotrophic bacteria principally from the genera Nitrosomonas and Nitrobacter. Nitrification occurs in percolating filters and activated sludge plants operated in a suitable manner. The first stage, sometimes referred to as "nitrosification", involves the oxidation of ammonium ions to nitrite and follows the general formula

$$NH_4^+ + 1.5\ O_2 \xrightarrow[\text{Nitrosomonas}]{} NO_2^- + 2H^+ + H_2O$$

In the second stage nitrite is oxidised to nitrate.

$$NO_2^- + 0.5O_2 \xrightarrow[\text{Nitrobacter}]{} NO_3^-$$

The overall nitrification process is described by the reaction

$$NH_4^- + 2O_2 \longrightarrow NO_3^- + 2H^+ + H_2O$$

Two important points are evident from this last reaction. Firstly, nitrification requires a considerable quantity of oxygen. Secondly, hydrogen ions are formed and hence the pH of the waste water will fall slightly during nitrification.

The settled sewage is effectively self buffering, but a fall of 0.2 of a pH unit is frequently observed at the onset of nitrification. In this autotrophic nitrification process, ammonia or nitrite provides the energy source, oxygen the electron acceptor, ammonia the nitrogen source and carbon dioxide the carbon source. The carbon dioxide is provided by the heterotrophic oxidation of carbonaceous nutrient, by reaction of the acid produced during nitrification with carbonate or bicarbonate present in the waste water or carbon dioxide in the air. Whereas for carbonaceous removal the oxygen requirement is roughly weight for weight with the nutrients oxidised, in the case of ammonia removal by nitrification approximately seven times as much oxygen is required to achieve the removal of the same quantity of nutrient.

Nitrification significantly increases the cost of sewage treatment since more air is required. Furthermore because these autotrophic organisms grow only slowly, longer retention periods are also required resulting in high capital costs. Nor does nitrification result in the production of an entirely acceptable sewage effluent. In areas where water re-use is practised the concentration of nitrate in river waters causes concern. There exists a limit on the concentration of nitrate in drinking water to avoid the occurrence of methaemoglobinaemia (so called "blue baby" syndrome). As a consequence denitrification is now practised after nitrification in some activated sludge treatment plants. In this anoxic heterotrophic bacterial process, nitrite and nitrate replace oxygen in the respiratory mechanism and gaseous nitrogen compounds are formed (nitrogen gas, nitrous and nitric oxides). However, this procedure is not part of conventional sewage treatment practice at present.

2.4 Secondary sedimentation

Both types of biological treatment require sedimentation to remove suspended matter from the oxidised effluent. Tanks similar to those normally employed for primary sedimentation are generally employed although at a higher loading of approximately 40 $m^3\ m^{-2}d^{-1}$ at 3DWF. Because of the lighter and more homogenous nature of secondary sludge simpler sludge scrapers are possible and scum removal is not necessary. The association of primary sedimentation tanks and a biological process for secondary treatment results in a sewage treatment works, as opposed to sewage farms where only land treatment was (is) employed. As an awareness of environmental pollution in addition to public health, developed in the 1950's and 60's, the term water pollution control works was introduced to describe sewage treatment works although this change of terminology was merely cosmetic. With the recognition of the importance of water re-use the term water reclamation works has found favour in some areas. Such works frequently apply additional tertiary treatment processes.

Sewage treatment results in the production of a final effluent suitable for discharge to the selected receiving water, and one or more sludges which may require treatment prior to disposal.

3. SLUDGE TREATMENT AND DISPOSAL

Sludge treatment and disposal is a facet of waste water treatment which is often given insufficient attention. Sludge treatment and disposal may account for 40% of the operating costs of a waste water treatment facility. Prior to treatment the sludges contain between 1 and 7% solids, which are usually highly putrescible and offensive. A wide range of treatment processes and disposal options have been used, although recently the cost of energy has reduced the numbers currently employed because of economic considerations.

The most convenient and economical method of disposal at any given site depends on a number of factors. Treatment of sludge is frequently influenced by the final disposal option selected. If sludge is to be disposed to sea from a works where the sludge may be pumped directly to the disposal vessel, then little treatment is required. Should the treatment works be close enough to the sea to make that type of disposal feasible, but not close enough to allow direct pumping to the disposal vessel, then economies in transport costs may be achieved by utilising some type of treatment process to thicken the sludge and reduce its water content prior to transport to the disposal vessel. If sludge is to be disposed of to land it is desirable to reduce transport costs, since the sludge will have to be spread over a wide area and, in the case of treatment works in urban locations, transported a significant distance to reach suitable land.

At present 67% of the sludge produced in the U.K. is disposed to land, 29% to sea and 4% is incinerated. Of the sludge disposed to land approximately two thirds is applied to agricultural and horticultural land and the remainder is used for land reclamation and land fill.

The processes available for sludge treatment include; thickening by stirring or flotation; digestion, aerobically or anaerobically; heat treatment; composting with domestic refuse; chemical conditioning with either organic or inorganic materials; dewatering, on drying beds, in filter presses, by vacuum filtration or centrifugation; heat drying; incineration in multiple hearth or fluidised bed furnaces and wet air oxidation. It is not feasible within this presentation to deal with all these processes in depth and the following is confined to the predominant sludge treatment process, anaerobic digestion and the most frequently utilised disposal option, that to agricultural land.

3.1 Anaerobic digestion

During anaerobic digestion the organic matter present in the sewage sludge is biologically converted to methane and carbon dioxide. The process is undertaken in an airtight reactor usually equipped with a floating gas collector. Sludge may be introduced continuously, but more frequently is added intermittently and the digestor operates as a "fill and draw" process. The methane produced is generally utilized for maintaining the process temperature, heating and power production by combustion in dual fuel engines, which use oil in the absence of methane.

Methane production is only significant at elevated temperatures, when 1 m^3 of methane at STP is produced for every 3 kg of BOD degraded. Digesters are characterized by the temperature at which they operate; those in which gas production is optimum at 35°C are described as 'mesophilic', whilst those in which gas production is optimum at 55°C are 'thermophilic'. These terms describe the temperature preferences of the bacteria undertaking the process.

Heat exchangers are used to transfer heat from the treated sludge to the influent sludge; the additional heat is provided by the combustion of methane. Digesters in the UK operate in the mesophilic range, since heat loss from thermophilic digesters would be unacceptable. To minimise heat loss digesters are frequently surrounded by earth banks to provide insulation. For efficient operation the digester requires a mixing system which may be mechanical or utilise the gas produced in the process to provide turbulence. A conventional anaerobic digester is illustrated in Figure 9. The result of anaerobic digestion is to reduce the volatile solids present in the original sludge by 50% and the total solids by 30%. In addition the unpleasant odour associated with the new sludge is drastically reduced. During the 20 to 40 days required for digestion the sludge is stabilised and emerges with a slightly tarry odour.

Traditionally anaerobic digestion has been considered a two stage process, a non-methanogenic stage followed by the methanogenic stage. The non-methanogenic stage has also been referred to as the acid forming stage since volatile fatty acids are the principal products. However, it is now recognised that the first stage may include as many as three steps. The first involves the hydrolysis of the fats, proteins and polysaccharides present in the sludge to produce long chain fatty acids, glycerol, short chain peptides, amino acids, monosaccharides and disaccharides. The second step (acid formation) involves the formation of a range of relatively low molecular weight materials including hydrogen, formic and acetic acids, other fatty acids, ketones and alcohol. It is now recognised that only hydrogen, formic acid and acetic acid can be utilised as substrates by the methanogenic bacteria. Thus in the third step compounds other than hydrogen, formic acid and acetic acid are converted by the obligatory hydrogen producing acetogenic (OHPA) bacteria. Some bacteria are able to undertake both steps 1 and 2 and produce hydrogen, formic acid and acetic acid which therefore do not require step 3. These stages are summarised in Figure 10.

Anaerobically digested sludge is frequently further dewatered in lagoons prior to disposal. Supernatant liquors are pumped from the surface of the lagoons to the head of the works for treatment.

Figure 9

Schematic diagram of an anaerobic digester

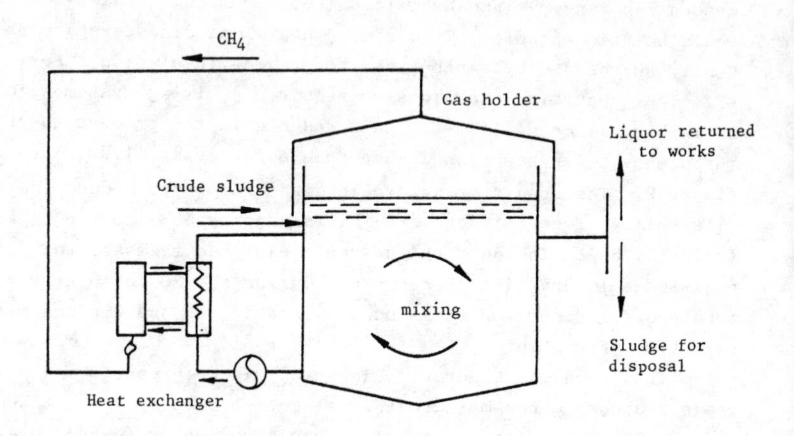

Figure 10

Biochemical transformations involved in anaerobic digestion

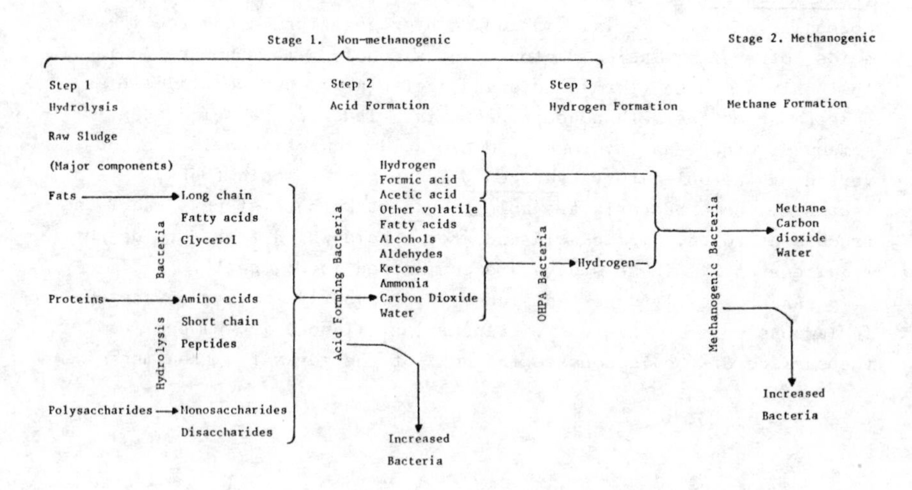

3.2 Disposal of sewage sludge to land

The practice of disposing sewage sludge to land has several potential benefits. Liquid sludges are conveniently applied to land by spray irrigation, from slurry tankers, while dried sludges may be spread and ploughed in. Some farmers value sewage sludge solely for its water content which may help to overcome irrigation problems during dry weather.

The application of sludge to land may help to slow down the decline in organic matter in soils under modern farming methods, leading to improvements in water holding capacity, porosity and aggregate stability. The main value of sludge as a fertiliser lies in its nitrogen and phosphorus contents. However, much of the nutrient content may be in organic forms and thus be unavailable to plants until mineralisation occurs. Although from an economic point of view it may be desirable to apply dried sludge to land, significant quantities of available forms of the nutrients may be lost during drying. Liquid digested sludge may contain up to 10% (W/W) of nitrogen, but only a fraction of this may be in available forms. For the purpose of calculating the available nitrogen in sludge, it is assumed that 85% of the total nitrogen in liquid digested sludge and 33% of that in dried sludge is available to crops during the growing season.

Since most agricultural soils are deficient in nutrients, fertilisers nearly always have to be added. However, sewage sludge is deficient in potassium, and therefore cannot fulfil complete fertiliser requirements. Furthermore, if all of the sewage sludge produced in the U.K. were to be applied to agricultural land, it would only provide 4.5% of the country's fertiliser requirement. Currently sewage sludge is applied to only 2% of agricultural land.

The potential hazards from the application of sludge to land are protozoal, viral, bacterial and other pathogens, which are present to the greatest extent in untreated sludges, persistent toxic organic compounds and toxic heavy metals.

Due to these hazards, salad or other crops which may be eaten raw should not be sown until one year after the application of treated sludge. If treated sludge is applied to pasture, animals should not be grazed until 3 weeks after the application. In the case of dairy cattle whose milk is not to be pasteurised the period of delay should be 5 weeks.

Heavy metals are of particular concern because they may be detrimental to crop growth or mobilised through the food chain. In many countries there exist guidelines designed to maintain the addition of sludge to land within safe limits. These limits specify maximum concentrations of several heavy metals which may be added on an annual basis over a period of up to 30 years. In the U.K. the limits of addition over 30 years are (in kg ha^{-1}) Cr and Pb 1000; B 50; As 10; Cd, Mo and Se 5; and Hg 2. The additions of Zn, Cu and Ni are governed by the Zinc Equivalent Concept, which appears to be the limiting factor for up to 75% of the sludges produced in the U.K. The zinc equivalent concept assumes that the phytotoxic effects of Zn, Cu and Ni are additive in the ratio 1:2:8. The recommended limit of addition of zinc equivalent is 560 kg ha^{-1}. The following is an example of the calculation of the zinc equivalent.

Assuming a sludge contains 1000 mg kg^{-1} Zn, 500 mg kg^{-1} Cu and 60 mg kg^{-1} Ni, then the zinc equivalent is given by:

$$\text{zinc equivalent} = (1 \times 1000) + (2 \times 500) + 8 \times 60$$
$$= 2480 \text{ mg kg}^{-1}$$

The recommended limit of application of the sludge (in tonnes of dry solids per hectare) over 30 years is given by:

$$\frac{\text{limit of addition of a particular metal (kg ha}^{-1}\text{) x 1000}}{\text{concentration of metal in sludge (mg kg-1)}}$$

which in the case of the above example is:

$$\frac{560 \times 1000}{2480} = 225 \text{ t ha}^{-1} \text{ over 30 years.}$$

This calculation is repeated, substituting values for the limits of addition of the other metals and their concentrations in the sludge, and the lowest value obtained is the maximum quantity of the sludge that may be applied.

The behaviour, fate and significance of organic micropollutants during sludge treatment and disposal are not well understood, and as a consequence no guidelines for the disposal of sludges contaminated with these materials exist. In the United States the Environmental Protection Agency has indicated 114 compounds of particular concern. Many of these materials fall into the following groups: polynuclear aromatic hydrocarbons, halogenated aliphatic and aromatic hydrocarbons, organochlorine pesticides, polychlorinated biphenyls and phthalate esters. Most of these materials are hydrophobic and not readily amenable to chemical or biological degradation. During sewage treatment they are intimately associated with the solids and therefore strongly concentrated into the sludges produced. They are therefore in the sludges to be disposed of. However knowledge of their significance is very limited. It appears that some chlorinated organics present in sewage sludge have been ingested (with soil) by grazing cows and that these materials have been passed to the milk associated with its lipid content.

4. REFERENCES

1. C.R. Curds and H.A. Hawkes (eds.) (1975) Ecological Aspects of Used-Water Treatment. Volume 1 - The Organisms and their Ecology, London: Academic Press.

2. Government of Great Britain (1970) Water Pollution Control Engineering. London: HMSO.

3. Government of Great Britain. Department of the Environment. (1970) Taken for Granted. Report of the Working Party on Sewage Disposal. London: HMSO.

4. Government of Great Britain. Department of the Environment/ National Water Council (1977) Report of the Working Party on the Disposal of Sewage Sludge to Land. Standing Technical Committee Report No.5. London:HMSO.

5. Metcalf and Eddy Inc. (1979) Wastewater Engineering: Treatment, Disposal, Reuse. New York: McGraw-Hill Inc.

6. The Open University (1975) Environmental Control and Public Health. Units 7-8 Water:Distribution, Drainage, Discharge and Disposal. Administrative Control. Milton Keynes: The Open University Press.

7. T.H.Y. Tebbutt (1973) Water Science and Technology. London: John Murray.

8. M. Winkler (1981) Biological Treatment of Wastewater. Chichester: Ellis Horwood Ltd.

6
The Chemistry of Metal Pollutants in Water

By D. P. H. Laxen
GRANT INSTITUTE OF GEOLOGY, UNIVERSITY OF EDINBURGH, WEST MAINS ROAD, EDINBURGH EH9 3JW, U.K.

Modern industrial society discharges vast quantities of metals into the environment. These metals become involved in complex biogeochemical cycles. Their ultimate destination (except on a geological time scale) is the sediments of the world's lakes and oceans. To reach the sediments the metals have to pass through the aquatic environment - streams, rivers, estuaries, oceans etc. Here the metals are available for uptake by a variety of aquatic organisms and, depending on a range of factors, toxic levels may be reached. In certain circumstances the metals will be passed up the food chain and become a major source of human exposure. This occurred in the now classic case of mercury poisoning in the population living around Minamata Bay in Japan in the 1950s. Industrial discharges of an organic mercury compound were being concentrated in the fish and shellfish, which were an important source of food in the area. Forty-three people died as a result of mercury poisoning before the pollution was stopped.

In order to develop sensible approaches to control pollution of the world's waters, it is necessary to understand what happens to the metals once they enter the environment. Considerable effort is required to unravel their complex biogeochemical cycles. Major advances have taken place over the last decade or so, largely since the advent of sophisticated analytical techniques, which allow the extremely low concentrations found in many waters to be accurately measured. As the science has developed, so it has become apparent that sample contamination can be a serious problem. It is now realised that, as a consequence, much of the early data is probably worthless.

There now exists a considerable body of data on metal pollutants in water. Much of this, though, refers to their total concentration. Unfortunately the total concentration of a metal provides little or no information about its bioavailability - the ability of a biological organism to take up the metal - nor its mobility - whether and for how long the metal will stay in solution or be removed to sediments. Indeed, it is now well established that to understand, and hence be able to predict and hopefully avoid, the adverse effects of metal pollutants in water, it is essential that we determine their physico-chemical forms (speciation), and the interactions between the various forms.

This paper will briefly describe the speciation of metal pollutants in different waters and the processes that determine the various species and their interactions. Examples will then be given of how the speciation determines the mobility of the metals, as well as how this can change as the metal moves from one type of water to another. This will be followed by several illustrations of how the speciation can affect the uptake of the metals by aquatic life. Finally some attention will be given to how water quality standards should in future reflect the speciation of the metals.

The Physico-chemical Forms of Metals in Water

Metals exist in a wide variety of possible forms as illustrated in Figure 1.

size µm: 0.001 — 0.01 — 0.1 — 1

soluble — colloidal — particulate

metal species	Free metal ions	Ion pairs; simple organic complexes; organo-metallic compounds	Complexes with high molecular weight organics	Metal species adsorbed on organic and inorganic colloids	Metal precipitates; mineral solids; living and dead organisms
example	Cd^{2+}	$ZnHCO_3^+$ Cu-glycinate $(CH_3)_4Pb$	Cu-humics	Pb-FeOOH Co-MnOOH Cu-humics Hg-clay Cd-humic/FeOOH	$Cu_2S(s)$ $Pb_3(OH)_2(CO_3)_2(s)$ Zn-feldspar Ni-organic

Figure 1 Possible metal forms in water classified according to size.

The simplest is the hydrated metal ion (a metal surrounded by water molecules). It has a positive charge, with a value of +2 for many metal pollutants.

These metal ions often form loose (electrostatic) associations with anions to create ion pairs, eg:

$$Me^{2+} + Cl^{-} \rightleftharpoons MeCl^{+} \quad (1)$$

where Me stands for a metal and Cl is the chloride anion. Ion pairs can be formed with charges ranging from positive through neutral to negative. The charge of the metal species may alter its behaviour. Considerable effort has therefore been devoted to defining the reactions causing ion pair formation. With this information it is possible to compute the distribution of a metal between ions and ion pairs. Unfortunately there is still considerable disagreement about the 'correct' values to use in the computation. Consequently models of the same system produce very different results, as shown in Table 1.

Metals form stronger bonds with organic molecules:

$$Me + \text{organic} \rightleftharpoons \text{Me-organic} \quad (2)$$

These range from simple complexes with amino-acids to those formed with humic substances. The latter are the predominant

Table 1 Computer modelling results for the inorganic speciation of soluble zinc in seawater [1,2]

Species	Year of study			
	1972	1975	1979	1981
Zn^{2+}	14.1	17.2	26.6	46
Zn-Cl*	79.5	10.6	47.0	35
$ZnSO_4^{o}$	1.7	3.5	4.3	4
$ZnOH^{+}$	0.9	0.2	4.4	12
$Zn(OH)_2^{o}$		62.8	0	
$ZnHCO_3^{+}$	3.8	0.7	1.0	3
$ZnCO_3^{o}$		5.0	16.7	

* Includes all Zn ion pairs with Cl

form of dissolved organic matter in natural waters. They have complicated and as yet poorly defined structures. Their molecular weights range from 300 to several 100,000s. The higher molecular weight humics can behave more like colloids - very small particles of 0.001-0.05 μm in size.

Metals which exist as ions, ion pairs and simple low molecular weight organic complexes can be considered to be soluble, or in true solution. Less common soluble forms of metal, but ones of considerable importance, are the organometallic compounds. In this case the metal is usually associated with one or more methyl groups. Examples are dimethylmercury $(CH_3)_2Hg$ and tetramethyl-lead, $(CH_3)_4Pb$. Organometallics can be discharged as such to the environment, or formed in situ.

Metal forms that are not soluble are called either colloidal (0.001-1.0 μm in size) or particulate (>1 μm). The metals may be associated with particles that are inorganic, organic or mixtures of both. Indeed, it is now known that inorganic particles in water are generally coated with a layer of adsorbed humic substances, so the mixed particle is probably more common.

Metals associate with particles as a result of a number of different processes. They can form part of the structure of the particle itself. For instance, clays and other minerals contain small amounts of metal within their crystal lattice. This metal is usually naturally present and cannot readily be detached from the solid. Metals also form precipitates if their concentration exceeds the solubility limit. Thus in certain waters lead can precipitate as lead carbonate, $PbCO_3(s)$. Organisms, such as algae and bacteria also take up metals, so creating another particulate metal form. Perhaps the most common way in which a metal becomes associated with particles is by adsorption. This process is described in more detail below.

The different metal species are usefully classified according to their size, as shown in Figure 1. It has become conventional when analysing waters to distinguish between 'dissolved' and 'particulate' metal using filtration with 0.45 μm filters. Examination of Figure 1 shows that this distinction is somewhat arbitrary. It does not really help define the different species and can even obscure the nature of the processes taking place. This problem does, nevertheless, highlight the fact that there is no analytical way to define the different metal species. Existing techniques can only group together broad classes of species.

Processes and Interactions between species

Several processes regulate the interactions between the different metal species: solubility control, adsorption, complexation, redox reactions, methylation and biological uptake. Each will be discussed briefly below, with the exception of biological uptake, which is discussed more fully in a later section. An important general point to make initially, though, is that metals differ in their response to these processes. It is therefore not possible to make sweeping generalisations about the chemistry of metal pollutants in water.

Solubility Control. Solubility, as it is defined here, represents the ability of a metal to stay in true solution. Metal salts dissolve until the solution is saturated or conversely in a saturated solution they precipitate:

$$Me^{2+} + 2OH^- \rightleftharpoons Me(OH)_2(s) \qquad (3)$$

In many cases the solubility is regulated by the concentration of OH^- ions. Thus solubility is often a function of the solution pH, as shown in Figure 2. The concentration of carbonate ions (CO_3^{2-}) is the other main control on solubility.

It is important to realise that the solubility reaction is dependent upon the concentration of simple metal ions (Me^{2+} in the above example). It has already been noted that metals also exist in true solution as ion pairs and organic complexes, equations 1 and 2. These reactions compete for the Me^{2+} and effectively increase the amount of soluble metal. Measured solubility can therefore be considerably higher than that predicted simply by use of equation 3. Metal concentrations in most waters are rarely high enough to exceed the solubility limit and cause their precipitation. One exception is the metal sulphides, which are of exceptionally low solubility. Sulphide rich waters occur where oxygen is absent and organic substances cause the reduction of sulphate

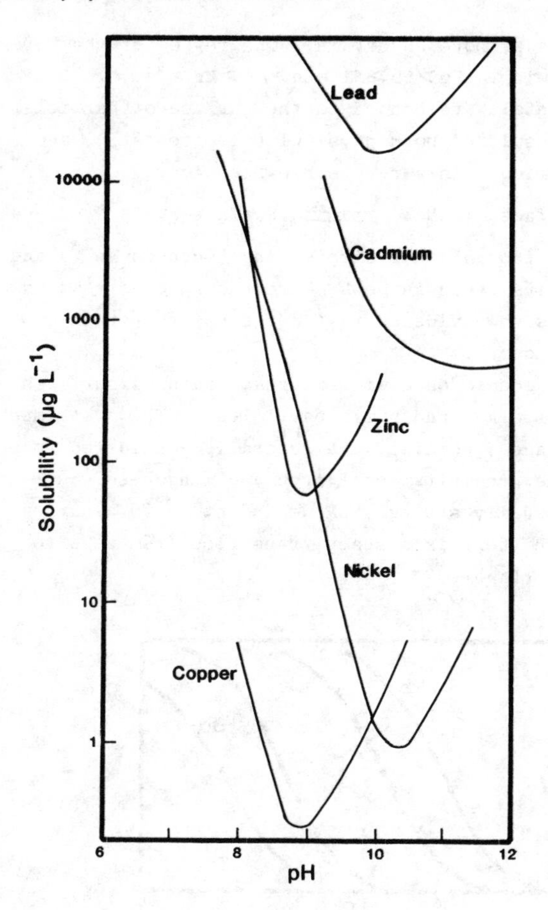

Figure 2 Theoretical metal hydroxide solubility as a function of pH.

to sulphide.

Complexation. Metals can form strong associations with organic substances (equation 2). This is generally held to occur by complexation, although metals might also be adsorbed onto the higher molecular weight humic substances. Indeed in a number of respects adsorption and complexation can become indistinguishable. Metals show considerable differences in the strengths of their reactions with organics. Copper and mercury form particularly strong complexes. The order of complexation for other metals is more uncertain. One review of the literature[4] gives the following: Hg > Cu > Ni > Zn > Co > Mn > Cd.

Adsorption. Metals are bonded to the surface of particles in water as a result of both physical (electrostatic) and chemical interations, in a reversible reaction:

$$\text{Surface} + \text{Me} \rightleftharpoons \text{Surface-Me} \qquad (4)$$

It is generally thought that simple ions (such as Me^{2+}) and possibly ion pairs (such as $MeOH^{+}$) are the species that adsorb. However there is now evidence that certain metal-organic complexes also adsorb.

Adsorption occurs on a wide range of natural surfaces, such as those of clays and humic particles. The most important adsorption surfaces, however, are generally considered to belong to the hydroxide precipitates of iron and manganese. Adsorption is usually strongly pH dependent, with a narrow range over which there is a sharp transition from zero to 100% adsorption (Figure 3).

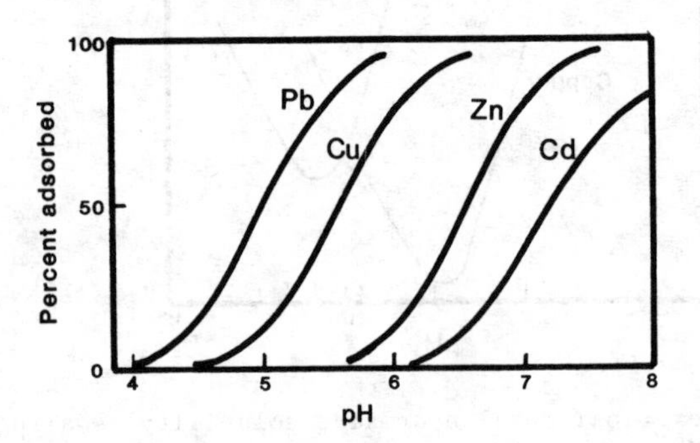

Figure 3 Adsorption of cadmium, copper, lead and zinc on amorphous iron hydroxide as a function of pH[5]. Reproduced with permission.

The tendency to adsorb follows the general order Hg > Pb > Cu > Zn > Cd > Ni > Co, although different studies find slightly different orders.

Redox Reactions. A number of metals are able to change their oxidation state in the environment. This occurs by a process of electron transfer, in simplied form:

$$\underset{\text{(reduced)}}{Me(II) - e^-} \rightleftharpoons \underset{\text{(oxidised)}}{Me(III)} \qquad (5)$$

Iron and manganese are two important metals that take part in redox reactions. Their reduced forms, Fe(II) and Mn(II), which are soluble, are found in waters and sediments where oxygen is absent. The oxidised forms, Fe(III) and Mn(IV), are, in contrast, highly insoluble. They therefore precipitate as hydroxides in oxygen rich waters. These precipitates produce fresh surfaces which can strongly adsorb metals.

A number of metals have vastly different properties dependent upon their oxidation state. For instance, chromium-(VI) is highly toxic, whereas in its reduced form, Cr(III), it is an essential element. As another example, the radioactive element plutonium is less readily adsorbed, and therefore more mobile in its oxidised forms, Pu(V and VI), than its reduced forms Pu(III and IV).

Methylation. A number of metal pollutants can be methylated in water or sediments. These include mercury, tin and possibly lead. In a simplified form the reaction for mercury methylation is:

$$CH_3 + Hg^{2+} \rightleftharpoons CH_3Hg^+ \qquad (6)$$

It is usually bacteria that cause the methylation and it is generally the more oxidised form of the metal that reacts.

The Influence of Speciation on Metal Mobility

As noted in the introduction, the forms in which a metal exists in a waterbody will influence its behaviour. That is, the form will determine how fast it moves through the system and whether it will stay in suspension or be removed to the sediments. A number of examples follow, which illustrate this point and also highlight the varied response of different metal pollutants.

Industrial Effluent Discharge. The first example focuses on the discharge of lead from a lead-acid battery manufacturer[6]. Waste waters from the industrial processes were being discharged to a river, after treatment to remove most of the lead. The lead concentration in the effluent was, nevertheless, still high, around 4000 $\mu g\ l^{-1}$. It appeared to be largely in the form of soluble ions and ion pairs. Some, though, occurred as particulate lead (>1 μm in size), probably in the form of lead sulphate formed during the treatment process (Figure 4).

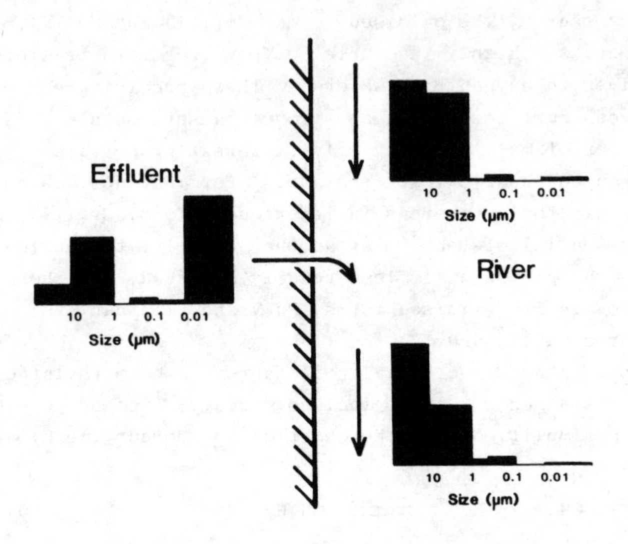

Figure 4 Percentage size distribution of lead species in an industrial effluent and in the receiving river.

In the river the speciation pattern was very different, at least on the occasion sampled. Most of the lead (98%) was associated with particles >1 μm in size (Figure 4). The discharge of industrial effluent caused a relatively small (30%) increase in the lead content of the river. But even though most of the lead in the effluent was soluble, once it mixed with the river water, it rapidly became attached to the larger particles, >12 μm in size. Particles of this size would rapidly settle once the river flow was reduced. (These samples were collected during a storm, and the fast flowing water was keeping particles in suspension.)

This example illustrates a more general point about the behaviour of lead. Namely, it is strongly adsorbed on particles. It will therefore readily be removed from the water as the particles settle. Indeed, studies have shown that

90-100% of the lead entering reservoirs is trapped in the sediments[7,8].

Sewage Effluent Discharge. A similar study to that cited above has been carried out for a sewage effluent polluted with cadmium[9]. About 60-70% of the cadmium entering the sewage treatment plant from an industrial source was being removed to the sewage sludges. The remainder, at a concentration of around 100 $\mu g\ l^{-1}$, was being discharged with the organic-rich sewage effluent to a small stream. The cadmium in the effluent appeared to be associated with organics, either as complexes with soluble organics or in association with colloidal organic matter (Figure 5).

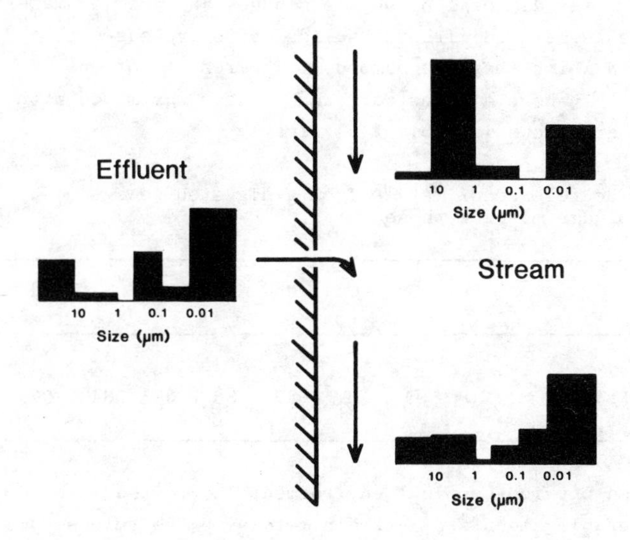

Figure 5 Percentage size distribution of cadmium species in a sewage effluent and in the receiving stream.

Within the stream the effluent caused a 29-fold increase in the cadmium concentration, to 9 $\mu g\ l^{-1}$, on the occasion sampled. Above the effluent input the cadmium was largely (61%) associated with particles $>$1 μm. Below the input most of the cadmium was found as soluble metal or in association with small (colloidal) particles. This cadmium would be unlikely to settle out quickly. It would therefore be transported fairly rapidly over long distances. In addition, by remaining in the water it would stay available for biological life.

The examples above illustrate two important controls over the transport of metals. Metals such as lead, which are strongly adsorbed on particles, can be easily removed from a waterbody and concentrated in sediments. Others such as cadmium are more likely to remain in solution, possibly complexed with organics. The latter are therefore considerably more mobile.

Release of Metals in Oceans. Considerable volumes of solid waste are dumped into the oceans each year. These arise mainly from the disposal of sewage sludges and dredged material. The metals associated with the solids may be released to the seawater in which they are dumped. The results of one experiment[10], where a digested sewage sludge was mixed with seawater, are shown in Table 2. Clearly

Table 2 The release of metals from a digested sewage sludge mixed with seawater.

Metal	Fe	Cr	Cu	Pb	Mn	Zn	Ni	Cd
Amount released (%)	0	4	9	35	36	39	64	96

the altered physico-chemical environment led to the release of considerable amounts of certain metals. The release depends upon how the metals were associated with the solids and the forms developed in the seawater. This example illustrates how metals incorporated into sludges (immobilised) in one environment (a sewage treatment plant) can be released in another (seawater).

Other Effects. The speciation of metals can alter the efficiency with which they are removed during industrial or sewage effluent treatment. In particular strong metal-organic complexes can retain the metals in solution, preventing their removal.

High concentrations of lead in drinking water can represent an important source of human exposure to this metal. They occur where there is lead plumbing and soft water. The lead pipes slowly release lead to the water running through them. The concentration reached depends upon the solubility of lead. This might be enhanced if the lead is complexed with organics or adsorbed onto small iron hydroxide particles (equations 2 and 4). Both these forms of lead have been found in drinking water samples[11].

It has already been mentioned that certain metals may be strongly adsorbed onto particles of iron and manganese hydroxide. In a number of lakes during the summer months the oxygen is removed from the bottom waters. In consequence any iron and manganese particles which enter these waters may be reduced (equation 5). Their reduced forms are soluble; therefore the particles dissolve, and in so doing release their adsorbed metal. This is one of the reasons why it is important to know what type of solid particles the metals are associated with.

Finally, the reversibility of the adsorption reaction (equation 4), coupled with its pH dependence, has important implications in relation to the growing concern about pollution from acid rain. The pH of a number of freshwaters is falling due to acid rain. This change in pH could lead to a release of adsorbed metals back into solution, where they will be more mobile and more toxic.

Metal Speciation and Biological Effects

A principal concern of pollution control strategies is the desire to avoid any harmful effects that metals may have on biological systems. In order to execute this responsibility it is essential that we understand how biological systems respond to metals in water. This response is crucially dependent upon the physico-chemical forms of the metal, as will become apparent in the following examples.

Uptake of metals in the case of small aquatic organisms, such as phytoplankton (uni-cellular plant life, eg. diatoms), is by diffusion across a biological membrane. Fish similarly transfer metals across membranes in their gills. This route for uptake is thought to be more important than transfer from ingested material in their gut. Only certain metal species are able to cross the membrane and it is these species that are bioavailable. There are however certain aquatic organisms which extract the nutrients they require by processing large amounts of particulate matter. These are the filter feeders such as oysters, clams, cockles etc. They often inhabit the surface of sediments and can extract metals from the sediment particles they ingest. In this case the bioavailability of the metals will depend on how they are associated with the particles.

Phytoplankton. These form an important food source for organisms higher up the food chain. Metal uptake by phytoplankton can lead to two adverse effects. Firstly, metals can be transferred up the food chain as the plankton are consumed by other organisms, fish etc. This effect, though, is probably not very important, as ingested metals are not highly bioavailable. Secondly, the metals can reach toxic concentrations in the phytoplankton. This might severely reduce an important food source and thereby have a profound effect upon the whole aquatic ecosystem.

Phytoplankton are known readily to accumulate metals. But it is only certain metal species that are taken up. These are generally thought to be simple metal ions. It is also likely that simple complexes and ion pairs - forms that can break up rapidly to form simple metal ions in the short time that they are close to the membrane - may be taken up. Strong complexes and adsorbed metals, on the other hand, are not available for uptake.

In one study of the toxicity of several metals to a marine diatom it was found that adding the strong organic complexing agent ethylenediaminetetraacetic acid (EDTA) reduced the toxicity of a range of metals[12]. Calculations suggested that the toxic effect was dependent solely upon the concentration of free metal ions and not the total metal concentration. When free metal concentrations (calculated by computer model, not measured) were compared the toxicity order was $Hg > Ag > Cu > Pb > Cd > Zn$.

In another study[13] of a mixed population of phytoplankton from a freshwater lake, it was found that the toxicity of a number of metals could be reduced if humic substances, sediments or the strong complexing agent nitrilotracetic acid (NTA) were added to the water. One exception was that mercury was more available in the presence of NTA. It only required a sediment concentration of 7.6 mg l^{-1} to reduce the bioavailability of the metals. The reduction in bioavailability followed the order Hg > Cu > Pb > Cd, with no effect found for Zn.

One further feature of phytoplankton worthy of note is their self-defence mechanism against metals. They can liberate organic substances which complex with the metals and thereby reduce their uptake. Finally, it should be mentioned that methyl-metals are more readily taken up by algae than other metal forms. In addition they are more toxic.

Fish. Fish, like phytoplankton have been found to respond only to certain metal forms. It should be emphasised at the outset, however, that this subject has only recently started to receive the attention it deserves. In consequence, there is still much to be learnt about how fish respond to the different metal species.

One recent study[14] of the toxicity of copper to cutthroat trout concluded that simple ions Cu^{2+} and the ion pairs $CuOH^{+}$, $Cu(OH)_2^{o}$ and $Cu_2(OH)_2^{2+}$ were the toxic forms, whereas $CuHCO_3^{+}$, $CuCO_3^{o}$ and $Cu(CO_3)_2^{2-}$ were not toxic. The precise details of this conclusion were reached using a computer programme to calculate the speciation. However, it was shown earlier that there is considerable uncertainty about the values to put into the programme and different researchers produce vastly different results for what is nominally an identical system (Table 1). This particular study did nevertheless establish that copper was more toxic in low alkalinity* waters, supporting the conclusion that ion pair formation with carbonate species reduces bioavailability (Figure 6).

It is rather better established that toxicity decreases as the hardness* of the water increases. This effect is also

---

* Alkalinity is essentially a measure of the carbonate and bicarbonate concentration expressed in units of $CaCO_3$ equivalence. Hardness is a measure of the calcium and magnesium concentration again expressed in $CaCO_3$ equivalence units.

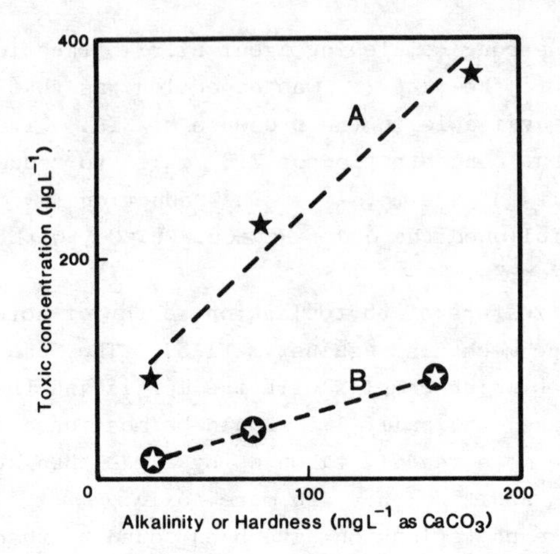

Figure 6 Relationship between the toxic concentration of total copper (96h LD50) and A) alkalinity for a constant high hardness, B) hardness for a constant low alkalinity.

shown in the study cited above (Figure 6). The calcium and magnesium ions causing the hardness do not change the copper speciation, but it is thought they interfere with the transfer of copper across the biological membrane.

The above study, along with a number of others reported in the literature,was conducted in a simplified system, in which only soluble ions and ion pairs existed. In real waters a considerable proportion of any copper will be complexed with organics or adsorbed on various solids. Several studies have shown that copper in these forms is largely unavailable to fish.

Filter Feeders. Filter feeders consume considerable volumes of particulate matter. Metals associated with the particles can be absorbed into the organism. The amount of metal taken up will depend, though, on how it is associated with the particles, ie. the form in which it occurs. In one set of experiments[15] the availability of silver, zinc and cobalt to clams was studied. The metals were adsorbed onto several artificial sediments before being presented to the clams in an aquarium with seawater. The sediments were iron hydroxide, organic particles (decaying marsh grass fragments), calcite (calcium carbonate), manganese oxide and biogenic calcium carbonate (ground up clam shells). The availability of the metals associated with these synthetic sediments is shown in Table 3.

Table 3 Availability to clams of metals adsorbed on synthetic sediments

Metal	Order of metal availability from sediment
Silver	Calcite > manganese oxide > iron hydroxide > biogenic $CaCO_3$ > organics
Zinc	Biogenic $CaCO_3$ > organics > calcite > manganese oxide > iron hydroxide
Cobalt	Biogenic $CaCO_3$ > calcite > organics > iron hydroxide > manganese oxide

It was found that the more easily the metal could be desorbed from the sediment, the more bioavailable it was.

A study has also been made of metal uptake by clams in their natural environment, estuaries[16]. Statistical techniques were used in this case, to help identify the important sources of metal uptake. For instance, it was found that the uptake of cadmium was not strongly related to the concentration of cadmium in the sediment. An important source of cadmium uptake was, apparently, 'available' cadmium in solution. For the other metals studied the sediments were the most important source. There was, however, some dependency upon the composition of the sediment. Thus association of lead with iron decreased lead uptake. On the other hand, the presence of humic substances in the sediment increased the uptake of cobalt and zinc.

These results emphasise that filter feeders respond in a variety of ways to the different metal forms. There are marked differences from one metal to another. In addition the response of different types of filter feeder is very diverse. In short the interactions between metals in both water and sediments and filter feeders are not straightforward. Considerable work remains to be done to unravel the complexity of effects.

Metal Speciation and Water Quality Control

It is useful to start this section by distinguishing between 'water quality standards', which are statutorily imposed limits, and 'water quality criteria', which imply desirable quality characteristics based on scientific evidence, without reference to political and legislative issues. Standards are therefore derived from criteria.

To date, water quality standards have been based almost entirely on total metal concentrations. However, to quote from a recent review of the speciation of trace metals in water[17], "It is inevitable that in the near future, water quality legislation for heavy metals will include statements relating to their speciation". This conclusion is amply justified by the examples given in this paper. Nowhere is it more important than here in Britain, where effluent discharge standards are based on the more flexible approach of water quality objectives[3]. The latter are set to safeguard the defined use of a particular stretch of water. Local emission standards are then set to ensure the quality objective is not exceeded. The whole procedure is outlined in Figure 7.

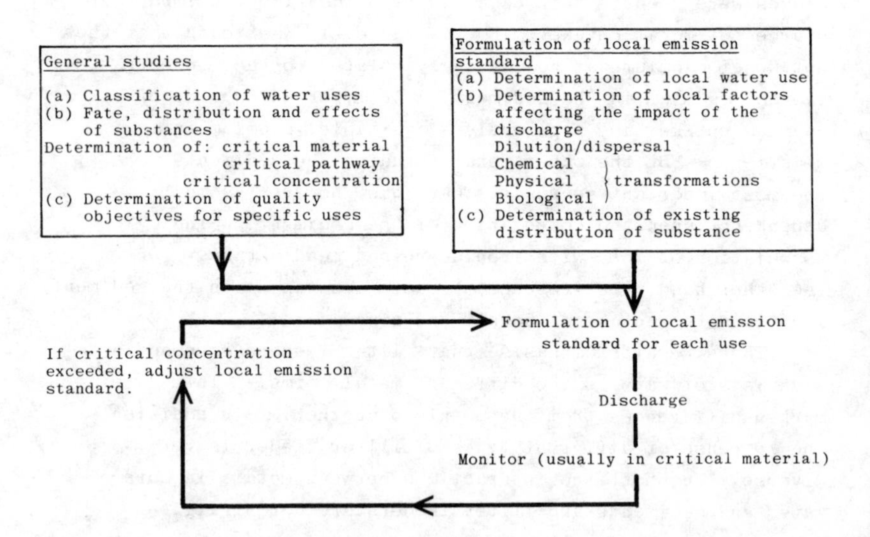

Figure 7 Procedure for use of water quality objectives to set local emission standards.

In theory, at least, it is of considerable importance that the fate of the discharge pollutant can be adequately defined. This inevitably requires attention to be paid to the speciation of the metal.

Steps are already being taken in the United States[18] of America to incorporate certain aspects of metal speciation into their water quality criteria. Criteria have been developed to protect all aquatic life from both acute (short term) and chronic (long term) effects. The former are expressed in terms of a concentration not to be exceeded at any time, whereas the latter guideline is expressed as a 24h average concentration limit. Criteria proposed for lead and cadmium are given in Table 4.

Table 4 Proposed water quality criteria for lead and cadmium[18].

	Criteria (μg L^{-1})				
	Freshwater				Saltwater
	Hardness (mg L^{-1} as $CaCO_3$)				
	20	50	100	200	
Lead					
24h average	0.09	0.75	3.8	20	insufficient data
Not to be exceeded	24	74	170	400	
Cadmium					
24h average	0.005	0.012	0.025	0.051	4.5
Not to be exceeded	0.6	1.5	3.0	6.3	59

The effects of water hardness on bioavailability have been incorporated into the guidelines (see earlier section on Fish). A distinction has also been made between the varied responses in freshwater and saltwater.

These criteria are, though, still expressed in terms of total metal concentration. In a symposium held in 1975 on speciation and toxicity of metals in natural waters[19], a number of participants argued that water quality criteria would be better expressed in terms of the dissolved metal concentration (ie. 0.45 um filterable). It should however be apparent from the previous discussion on metal forms that this would not be a wholly satisfactory way of dealing with metal toxicity. It would though represent a temporary expedient, recognising that bound metals are not particularly bioavailable.

More recently it has been suggested[17] that criteria might be based on metals measured by anodic stripping voltammetry in sea water and by ion-exchange resins in freshwaters. These techniques should measure concentrations of metal ions, most ion pairs and metal complexes that break-up rapidly, ie. the forms that are in theory available for transport across membranes (this was discussed under the section on Phytoplankton).

Several problems may arise though from basing criteria and standards on a measureable fraction of the total metal. Firstly, this approach does not take account of transformations of metals from one form to another. Secondly, it does not allow for the fact that the unmeasured portion may be bound to particulates in a form that is available to filter feeders.

It should be apparent from this brief introduction to the subject, that whilst aspects of speciation need to be incorporated into water quality criteria and standards, there is as yet no clear way to do this. Considerably more effort needs to be devoted to this topic, if we are to develop realistic and rational means of protecting our environment from metal pollutants.

References

1. T.M. Florence and G.E. Batley, CRC Critical Rev. Analyt. Chem., 1980, 9, 219.
2. D.R. Turner, M. Whitfield and A.G. Dickson, Geochim. Cosmochim. Acta, 1981, 45, 855.
3. R.M. Harrison and D.P.H. Laxen, "Lead Pollution - Causes and Control", Chapman & Hall, London, 1981
4. R.F.C. Mantoura, A. Dickson and J.P. Riley, Estuar. Coast. Mar. Sci., 1978, 6, 387.
5. M.M. Benjamin and J.O. Leckie, J. Colloid Interface Sci., 1981, 79, 209.

[6] D.P.H. Laxen and R.M. Harrison, Water Res., in press.

[7] C.W. Randall, T.J. Grizzard, D.R. Helsel and R.C. Hoehn, Proc. Int. Conf. "Heavy Metals in the Environment", Amsterdam, Sept. 1981, CEP Consultants Ltd., Edinburgh.

[8] J.P.C. Harding and B.A. Whitton, Water Res., 1978, 12, 307.

[9] D.P.H. Laxen and R.M. Harrison, Water Res., 1981, 15, 1053.

[10] N. Rohatgi and K.Y. Chen, J. Water Pollut, Contr. Fed., 1975, 47, 2298.

[11] R.M. Harrison and D.P.H. Laxen, Nature, 1978, 275, 738.

[12] G.S. Canterford and D.R. Canterford, J. Mar. Biol. Assoc. U.K., 1980, 60, 227.

[13] D. Hongve, O.K. Skogheim, A. Hindar and H. Abrahamsen, Bull. Environ. Contam. Toxicol., 1980, 25, 594.

[14] C. Chakoumakos, R.C. Russo and R.V. Thurston, Environ. Sci. Technol., 1979, 13, 213.

[15] S.M. Luoma and E.A. Jenne, In "Biological Implications of Metals in the Environment", Available as CONF-750929, U.S. NTIS, Springield, VA., 1977, p213.

[16] S.N. Luoma and G.W. Bryan, Estuar. Coast. Shelf Sci., 1982, 15, 95.

[17] T.M. Florence, Talanta, 1982, 29, 345.

[18] U.S. Environmental Protection Agency, Federal Register, 1980, 45, 79318.

[19] R.W. Andrew, P.V. Hodson and D.E. Konasewich (Eds.), Proc. Wkshp. "Toxicity to Biota of Metal Forms in Natural Water", Duluth, Minn., Oct. 1975., International Joint Commission, Great Lakes Research Advisory Board, Windsor, Ont.

7
Effects of Pollutants in the Aquatic Environment

By H. A. Hawkes
DEPARTMENT OF ENVIRONMENTAL SCIENCES, UNIVERSITY OF LANCASTER, LANCASTER LA1 4YQ, U.K.

Introduction

The biosphere is that depth of the earth's surface where the incoming light energy from the sun is trapped by the functioning of ecosystems involving the interaction of the different living (biotic) components--Producers, Consumers and Decomposers--with the non-living (abiotic) components to produce a cyclic interchange of materials between them (Fig. 1). With the exception of a few micro-organisms (chemolithotrophs) which are capable of utilising the energy from exothermic inorganic chemical reactions, all life on earth, including Man, is dependent directly or indirectly on the functioning of these ecosystems. The maintenance of a viable biosphere (Conservation) is therefore in Man's long-term interest. Approximately 70 percent of the globe's surface is covered with water, much of which is inhabited to a considerable depth. The aquatic environment (Hydrosphere) therefore represents globally a significant proportion of the biosphere. Of the total water in the hydrologic cycle over 97 percent is sea; most of the remainder is frozen assets in the form of polar ice caps and glaciers. Freshwater lakes and rivers represent 0.009 and 0.0001 percent of the total global water. The significance of inland freshwaters to Man however is out of all proportion to this figure. They

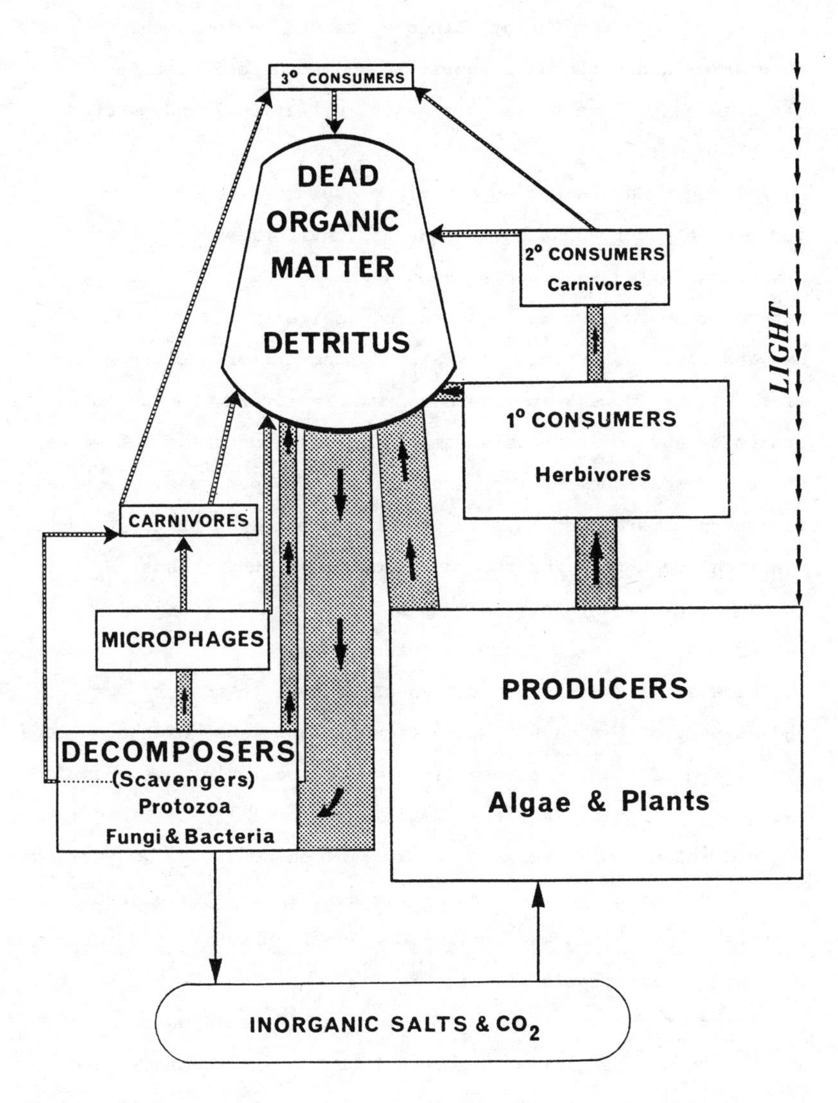

Figure1. Diagrammatic Representation of an Ecosystem.

provide him with water and carry away his effluents;they serve as sources of food and energy, lines of communication,local boundaries and national borders. In advanced societies they may also contribute to satisfy Man's recreational and general aesthetic needs.

Although Man is not an aquatic organism he has become increasingly dependent on the aquatic environment to satisfy his technological and social needs. Water bodies form both a natural resource for Man's domestic, agricultural and industrial use and also part of his natural environment for his other uses listed above. Freshwaters and especially rivers, because of Man's proximity and their relative small size,are generally subjected to greater environmental pressures than the open seas. There is for example more concern shown for the protection of the trout in the local beck than for the phytoplankton in the oceans. However in the longterm interest of Man's future the protection of the latter globally may be more important.

Because of Man's various uses of water,both as a resource and as part of his natural environment,and because of the various types of pollution —outlined later— water pollution has been variously defined. Earlier definitions,especially by biologists, defined pollution in terms of ecological damage. In cases of legal action however ecological change was not, in itself, always acceptable as evidence of pollution. A more practical definition more widely acceptable relates to the impairment of the natural qualities of the water so as to affect Man's legitimate uses.[1,2] A more recent definition [3] covers both the environmental and resource aspects-"The introduction by Man into the environment of substances or energy liable to cause hazards to human health, harm to living resources and ecological systems,damage to structure or amenity or interference with the legitimate uses of the environment". In the context of the foregoing discussion on Man's uses of the aquatic environment, the following definition of water pollution is proposed for the purposes of this chapter-

> "The discharge of organisms, substances or energy to water, resulting from Man's activities,which impair it's legitimate uses as a natural resource or as a natural environment".

Having developed the concept of pollution, we shall now review the several types of pollution by considering the effects of different pollutants in the aquatic environment. An appreciation of these different effects is essential in controlling pollution in order to apply appropriate treatment and monitoring methods.

Factors Affecting Organisms in the Natural Aquatic Environment

Pollutants in the aquatic environment may be substances or agents foreign to the natural environment e.g. synthetic pesticides. Pollution may also be caused by the occurrence of abnormal levels of natural factors e.g. high phosphate concentrations,low oxygen concentrations.To understand the effects of pollutants therefore it is first necessary to appreciate the relevant natural factors affecting organisms in the aquatic environment.

Water as a Medium for Life. Several physical properties of water are significant in characterising the aquatic environment and the forms of life which inhabit it. Compared with the aerial and sub-aerial environments water provides a much denser medium. The resultant buoyancy makes possible modes of life not possible in other media, such as plankton, which drift suspended in the water column, and filter-feeding. Of considerable ecological significance is the elementary fact that the density of water is greatest at 4°C,above and below which temperature increases or decreases result in a decrease in density,this effect being greatest at high temperatures. This phenomenon results in the stratification of lakes,ice formation on the surface of waters and the induction of convection currents. The surface tension of water makes possible a specialised community of organisms living at the air-water interface known as the neuston.

Marine organisms have protoplasm which is isotonic with their external medium. Because of the lower salt concentration of freshwaters,the protoplasm of freshwater organisms has a higher osmotic pressure than the external medium giving rise to problems of osmoregulation.The salinity of freshwaters is therefore of significance in influencing the distribution of organisms.

Chemically, natural water in the aquatic environment is not H_2O but a dilute solution of biogenic materials including nutrient salts and gases. In nature its composition differs appreciably in different waters. Thus the definition of water purity in purely chemical terms is not possible.Because of the great solvent properties of water,without involving chemical reaction,all substances required to support life,both macronutrients and micronutrients,are soluble in water.Of the macronutrients nitrogen and phosphorus are important limiting factors in the aquatic environment. Calcium is one of the biggest variables in freshwaters, characteristic communities being associated with soft and hard waters. Some organisms have specific requirements for certain elements;silicon is important in diatom ecology. Silicon is present in small amounts in solution as orthosilicates but is mostly present as the colloidal or particulate form. As mentioned earlier, salinity,the concentration of all ionic constituents—carbonates,sulphates and halides of sodium,potassium and magnesium, is significant in determining the osmotic pressure of the water.

The two gases of ecological significance—carbon dioxide and oxygen—behave differently in solution. Carbon dioxide is very soluble in water entering into chemical reaction to establish an equilibrium which is greatly influenced by pH:

$$\begin{array}{ccc} \text{Gas} & & \text{Dissolved} \\ CO_2 & \rightleftharpoons & CO_2 \end{array}$$

$$CO_2 + H_2O \rightleftharpoons H_2CO_3 \rightleftharpoons H^+ + HCO_3^-$$

$$HCO_3^= \rightleftharpoons H^+ + CO_3^-$$

Oxygen, in contrast, is poorly soluble in water and its solubility is greatly influenced by temperature. Dissolved oxygen therefore is a major limiting factor in the aquatic environment. For example compared with 210 ml of oxygen available to aerial organisms in each litre of air there is, even at saturation, in one litre of water approx. 10.2 ml. at $0^{\circ}C$, 7.9 at $10^{\circ}C$ and 6.3 at $20^{\circ}C$.

Aquatic Ecosystems. Although the above factors are important in determining the different aquatic ecosystems, e.g. marine and freshwater, other factors are responsible for the distribution of organisms within them. In freshwaters a major factor is current. Lentic waters—bodies of standing water such as ponds and lakes — and lotic waters — flowing waters such as streams and rivers — are distinguished. Different environmental factors are important in the two types of habitat. In lentic waters the depth profile in relation to light penetration and thermal stratification are major ecological parameters. In lotic waters the current velocity and the resultant nature of the substratum are important factors influencing the the distribution of organisms. A rapid current over a stony eroding subsratum supports quite a different community than a sluggish current over a depositing substratum, even in the same stretch of river. On the basis of these differences different river zones may be recognized, the major divisions of which are the upland stretches- Rhithron— and lowland stretches—Potamon.

All natural aquatic ecosystems involve the cycling of materials through Producer, Consumer and Decomposer populations. The presence therefore of organic matter and of those organisms responsible for its degradation such as saprobic micro-organisms and scavenging invertebrates are part of the natural system. In lentic freshwaters e.g. lakes much of the material cycles within the system— a closed ecosystem— although there are both aerial and aquatic imputs of allochthonous materials and a loss via the outflowing stream. In lotic systems— rivers— especially in the rhithron zone, because of the current transporting materials downstream, most of the materials are of allochthonous origin being derived from the terrestrial ecosystems. Lotic ecosystems may therefore be regarded as Open Ecosystems (Fig.2).

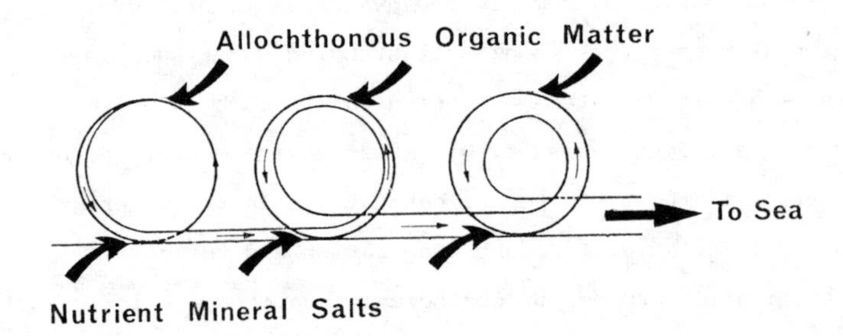

Figure 2 Diagrammatic Representation of a River Open Ecosystem.

Closed lentic ecosystems are therefore relatively independent of allochthonous imputs whereas open lotic systems depend upon it. As a result lotic systems tend to be more sensitive to changes in inputs—such as effluents— and respond sharply. Lentic systems are less sensitive and generally respond more slowly although the eventual change may be as great and more longlasting.

Classification of Pollution

Different types of pollution have been variously classified [3,4]. Appropriate to our present topic we shall here, for the sake of presentation, classify them according to the ecological factor most affected although it will be appreciated that many discharges may affect more than one factor simultaneously. Table 1 presents such a classification and includes examples of pollutants and their sources. The different types of pollution are then discussed in more detail in the subsequent sections.

Table 1. Classification of Pollution based on Ecological Factors

Type	Factor	Examples of Sources
PHYSICAL	Suspended Solids	Mining, Quarrying, China clay, Pulping
	Turbidity	Most Effluents-sewage and industrial
	Colour	Dye manufacture, Textiles, Tanneries Paint manufacture.
	Surface Properties	Detergents-domestic and industrial uses, Oils, Engineering, Synthetic Rubber manufacture.
	Temperature	Industrial cooling waters - Electricity generation, Steel mills.
	Radioactivity	Processing nuclear fuels, Electricity generation.
CHEMICAL	Salinity	Coal and Salt Mining, Oil Wells.
	pH	Mine waters, most manufacturing industry effluents, Battery manufacture
	Toxicity	Tanneries, Wood pulping, Manufacture of Dyes, Explosives, Man-made fibres, Paints, Pesticides, Pharmaceutics, Photofilm, Plastics, Synthetic rubber, Processing and mining of metals.
	Deoxygenation	Mine waters.
BIOTIC	Nutritional-	
	Organic	Processing of Biological Materials-Foods, Drinks, Antibiotics, Paper, Rubber, Textiles, Leather, Glue, Tar. Sewage effluents, Farm effluents.
	Inorganic	Agricultural land drainage, Sewage effluents, Fertiliser manufacture.
	Intrusive Organisms	Sewage effluents, Processing of Meat, Tanneries.

Physical Pollution

The discharge of substances which are neither toxic nor deoxygenating may affect the physical characteristics of the water itself or of the substratum and thereby adversely affect the aquatic biota.

Suspended Solids, Turbidity and Colour. These factors all tend to affect the aquatic environment in a similar manner by reducing light penetration and thereby plant life generally, with consequent secondary effects on food chains. The sparsity of benthic organisms in certain streams in Cornwall has been attributed to the turbid conditions caused by china clay wastes which suppress algal growth in the streams[5]. The direct effects of suspended solids on organisms is difficult to assess since the deposition of the solids on the stream bed is a complicating factor. Observations on a rapid-flowing river in S.Wales, which was in a stage of recovery from pollution by effluents from a series of coal washeries following the installation of a more efficient treatment plant, showed that over a period of one year, during which the river was subjected to intermittent discharges of supended coal material, a benthic invertebrate fauna including *Gammarus pulex* (Freshwater shrimp) and several species of mayflies became established. It was observed that amongst the stones in the river bed were dead tree-leaves which provided food for the invertebrates which were not therefore dependent on algae. Animals which feed by straining materials from the water such as the net-spinning caddis *Hydropsyche* were notably absent, whereas the predatory caddis *Rhyacophila* was common.

Turbidity may affect the outcome of predator-prey interaction by reducing the ability of predators which hunt by sight. The abundance of leeches in highly turbid waters was attributed to reduced predation by fish. Fish which feed by sight and anglers who make use of this are both affected by turbidity. The suppression of benthic communities results in the reduced availability of fish food. In the china clay polluted rivers of Cornwall trout fry were absent but 2-yr.old adult trout were present but at approx. one-seventh the population found in the non-polluted stretches. It was concluded that fish were not breeding in the polluted waters but migrated into them from the non-polluted stretches.The individual fish were of similar size and condition to those in the unpolluted water suggesting that there was sufficient food to support the restricted population. Examination of the gut contents revealed that the fish in the polluted waters took a higher proportion of food of terrestrial origin. The drift in the polluted waters consisted of a higher proportion of terrestrial animals. Some of the fish taken from the affected stretches were found to have gills with thickening of the epithelial cells and fusion of adjacent lamellae.This effect had been observed in experimental trout subjected to long periods of immersion in suspensions of kaolin and diatomaceous earth, but other fish under the same conditions had normal gills. These pathological changes could affect the respiratory efficiency of the gills. These tests of 185 days duration showed that at 30mg l^{-1} the death rate was negligible,at90mg l^{-1} a few fish died but at 270 and 810mg l^{-1} a high proportion of the fish died[6]. Other work showed that the nature of the suspended solids was more important than their concentration in determining the direct effect on fish. 200mg l^{-1} wood fibre caused substantial mortality when coal dust at this concentration had no effect[7].At higher concentrations there was no relationship between survival period and concentration;it was considered that increased susceptibility to disease was an important factor.

Many natural rivers carry heavy loads of suspended matter at times of spate . Although fish may not be killed during several hours or even days exposure to several thousand mg l^{-1} solids, such conditions should be prevented where good fisheries are to be maintained [8] .

More important ecologically than the direct effects of suspended matter in the water is the effect on the benthos when it settles and smothers the natural substratum.The deposition on a natural stony substratum of inert materials such as coal dust, gravel washings,metal mining wastes or flocculated metal compounds results in a general reduction in the benthos in terms of both species and populations,thus affecting fish food. Organic solids however such as present in sewage effluents and effluents from textile, hardboard and paper manufacture form organic sludges on the river bed. In these the natural benthic community of an eroding substratum is replaced by a silt community with such species as *Chironomus riparius* (Blood worms) and *Asellus aquaticus* (Water hog louse) which often develop high population densities to provide an abundant food supply for fish. The most sensitive freshwater habit in respect of suspended solids is probably the spawning grounds of trout and salmon where the slightest turbidity in the water or deposition of solids may cause the fish to avoid the area or result in failure of the eggs to develop successfully after they have been laid[8].

Apart from these special requirements for salmonid breeding the E.I.F.A.C.* tentative criteria for chemically inert solids in waters which are otherwise satisfactory for fisheries are summarised in Table 2.

* European Inland Fisheries Advisory Commission.

Table 2 E.I.F.A.C. Criteria for inert suspended solids in respect of Freshwater Fisheries. Based on abstracted data [8].

Suspended solids concentration mg l^{-1}	Effect
<25	No harmful effects
25 — 80	Good — moderate fisheries possible although yield may be somewhat reduced.
80 — 400	Unlikely to support good fishery although some fisheries found at lower concentrations in this range.
> 400	At best only poor fisheries found.

The presence of suspended solids in water detracts from its use for most purposes and adds to the cost of treatment.

Colour in river water may result from the discharge of coloured effluents, e.g. dye manufacture or textiles, or it may appear as the result of the interaction of different effluents in the river. Vegetable tanning effluents and ferruginous waters produce a deep green to inky blue colour. The ecological effect of colour will theoretically depend upon its light absorptive properties in relation to the spectral requirements of the algae and plants. Many rivers, such as those draining peat, are naturally highly coloured and yet support a typical biota including trout. Although evidently polluted, and of reduced quality for many purposes, the ecological effects of colour itself are usually minimal compared with other factors.

Surface Active Agents. The advent of synthetic detergents for domestic use in the years following World War II exemplifies the impact of technology on the environment. Although natural soaps affect the physical properties of the water in a similar way to synthetic detergents, they are rapidly degraded whereas the earlier detergents were resistant to biological degradation. The development of the biologically soft detergents exemplifies the use of technology to minimize the technological impact on the environment.

The presence of surface active materials in sewage effluents suppresses the rate of re-aeration of the receiving stream and thereby delays the self-purification, this effect being greater in sluggish rivers than under quiescent conditions or in torrential rapid-flowing waters.Theoretically since many aquatic insects are associated at some stage of their life hitory with the air-water interface, the effect of surface active agents on the surface tension might be expected to affect such insects. Also the presence of foam on the surface of the water would affect the emergence and egg-laying activities of some insects.

Besides the physical effects outlined above,detergents are toxic to fish. This has been attributed to the reduced surface tension at the surface of the gill,resulting in symptoms characteristic of asphyxia leading to death. Other workers however have attributed the toxicity to chemical effects. Generally the soft detergents have been found more toxic than the older hard forms. Many packaged detergents contain a high proportion (up to 50 percent.) of polyphosphates as builders. The widespread use of these products contributes substantially to the phosphate levels in the aquatic environment which are of significance in relation to eutrophication as discussed later.

Temperature. The ecological effects of elevated temperatures in the aquatic environment are very complex. All aquatic organisms have a fairly well defined temperature tolerance range and this determines their distribution. The simplest effect of elevated temperatures is to eliminate the cold-water temperature sensitive species — stenotherms — from the heated waters. Such species however, e.g. the flatworm Crenobia alpina and stoneflies, live in small upland streams which rarely receive thermal discharges. The temperature in such streams may however be elevated by Man's other activities such as deforestation. In N. America, where this resulted in the elevation of the stream temperature by only a few degrees, trout populations were seriously affected[9]. In the larger lowland rivers which are used for cooling and therefore receive thermal discharges, the biota is naturally eurythermal being conditioned to a wider range of temperatures and is therefore less affected directly by increased temperatures. Such discharges would appear to have no drastic effect providing the river water temperature does not exceed 30°C. Temperature however has more subtle indirect effects on aquatic ecosystems.

The life-cycles of aquatic insects are conditioned by seasonal temperature fluctuations. By elevating the winter temperature of the water the timing of the life-cycles could be changed so that the emergence of the spring mating flight could occur earlier when aerial conditions were unsuitable. A study of the effects of heated water discharges on a good quality river (R.Severn at Ironbridge), however, showed that any such effect was masked by other variabilities [10]. A more fundamental effect of elevated temperatures is that on productivity. Studies on the effects of increased temperatures of some Polish lakes (Konin Lakes) following their use as cooling water showed that the biomass of non-predatory plankters — rotifers and cladocera — increased by a factor of six. In the most heated lake there was a reduction in the abundance of bottom-dwelling chironomids, but oligochaete worms increased. It was considered that the heating of the water reduced the value of the lake bottom as a feeding ground for fish and it was found that the bottom-feeding fish were most affected. It was concluded generally that elevated temperatures would not adversely affect fish food organisms within the range the fish themselves could tolerate.

The direct effect of temperature on the growth rate of organisms is determined by the proportion of the increased metabolic activity that is channelled into synthesis compared with cellular respiration. With limited food or suppressed feeding activity, increased temperatures could result in reduced growth rates. With unlimited food and unrestricted feeding, increased growth rates would probably result. Heated industrial cooling waters have been used in the culture of marine and freshwater fish and significantly increased growth rates have been achieved. The terms "Thermal Pollution" and "Thermal Enrichment" have been used to describe warming up of natural waters by thermal discharges; a more objective term is calefaction, the detrimental or beneficial effects of which depend upon the uses of the water.

Indirect effects of calefaction are equally important. The most serious is probably the accentuation of the oxygen depletion resulting from organic pollution. The effect on toxicity is more complex; although the rate at which fish die in toxic solutions is increased at higher temperatures, the threshold concentration of a substance may not be reduced and may even be increased, due to enhanced detoxification processes at higher temperatures, e.g. phenol[12]. Temperature also affects the development of parasites and pathogenic bacteria and may also affect the resistance of fish to disease.

For the protection of inland fisheries it has been recommended that for salmon and trout waters the upper permissible temperature during the warmest season of the year is 20-21°C. For other fish—cyprinids—a maximum of 28°C is implied[8]. For spawning and early developmental stages lower temperatures are necessary, differing according to species [8]. An increase of only 2 centigrade degrees above normal during the breeding season is detrimental to the breeding of burbot (Lota Lota) and a 5 — 6 centigrade degree rise above seasonal normal could be detrimental to salmonids.

Radioactivity. Although in normal circumstances only low-level activity enters inland waters, it is possible for stream ecosystems to concentrate these via food chains by factors greater than those achieved by dilution in disposal. This also applies to food chains leading to fish, Cs137, Sr85 and P32 probably being the most important isotopes in this respect. Many algae have been shown to be capable of concentrating radioactive materials from low levels in the water. There is a danger of radioactive materials, concentrated by algal blooms from raw waters, being released into supply water. The deposition of radioactive algae in estuaries as a radioactive sludge is also possible. These possibilities should be considered when determining the permissible levels of activity for discharges to the aquatic environment.

Chemical Pollution

Some effluents change the concentration of the natural chemical components of the water causing un-natural levels of naturally occurring substances. Other effluents, mostly industrial, introduce substances quite foreign to the natural aquatic environment, many of which are detrimental to aquatic organisms and water quality generally. This section deals with the effects of these changes in the chemistry of the water, excluding nutritional effects which are included in biotic effects discussed in the next section.

Salinity. Freshwater organisms, which have become adapted to living in waters of lower osmotic pressures than their body fluids, may be adversely affected by increases in the osmotic pressure resulting from saline discharges. Probably a more important ecological effect of such discharges however is the change in the proportional ionic composition. In natural waters salts of Na, Ca, K and Mg are present in such proportions that their individual toxicities are mutually counteracted or antagonized. Discharges which increase the concentration of one ion may result in a toxic solution. Discharges which result in fairly stable saline conditions, such as those from coal mines, salt mines and oil wells, will encourage the development of a replacement community of brackish water species. The brine shrimp — *Artemia salina* —, the salt fly — *Ephydra riviparia*, certain rotifers and diatoms have been found to be characteristic of such communities.

Increased salinities in inland rivers permit the invasion of estuarine species. Associated with a saline discharge into a tributary of the R. Trent in Staffordshire, *Gammarus tigrinus* — an estuarine species — has progressively replaced *G. pulux* as the dominant species in the tributary and in the river downstream. In the R. Severn *Corophium curvispinum* has become established at least as far as Bewdley. Intermittent discharges of saline effluents may however suppress the natural community without creating stable saline conditions to permit the establishment of a replacement one.

The chloride concentration of natural freshwater differs considerably. Besides affecting the biota, high salinities affect the quality of the water for public supply and agricultural irrigation. The W.H.O. limit for drinking water is 200 mg Cl l^{-1}; the use of the formally permissible level under certain conditions of 600 mg l^{-1} is now discouraged because of the health significance of Na[13]. Salinity is not considered a major pollutant in rivers in connection with freshwater fisheries.

pH. The effects of pH itself are difficult to establish since the substances causing the pH may themselves have direct effects and many industrial effluents affecting the pH of the water also contain other pollutants e.g. plating wastes. Within the range pH 5.0—9.0 pH probably has no significant direct effect on most species, although some taxa are more restricted e.g. snails are usually found in waters with pH above 7.0 . In South Africa the effects of discharges of sulphuric acid produced by the oxidation of pyrites exposed by coal and gold mining activities have been studied[14]. The pH in the receiving streams was for the most part below 4.8, reaching 2.9 in one stream. The bottom fauna was severely restricted to chironomid larvae and certain oribatid mites. Mayflies, caddisflies, beetles, Simulium flies and certain non-biting midges were all eliminated. Certain algae flourished and in the pools Sphagnum moss covered the stream bed. In a survey of acid waters in England[15] the flagellate protozoon Euglena mutabilis was found to be the most common and abundant species; in relatively slow flowing acid mine waters it has been observed as a green film covering most of the water surface. Thus under conditions of very low pH a restricted replacement community became established as with saline pollution. In the absence of competition and possible predation by fish, the surviving species establish large populations. Again with intermittent discharges the fluctuating pH permits neither the natural nor the replacement community to exist.

Probably more important ecologically than the direct effect of pH is the indirect effect in influencing the toxicity of many substances,especially in which the toxicity depends upon the degree of dissociation. A nickelocyanide complex was found to be 500 times more toxic to fish at pH 7.0 than at 8.0 [8], whereas ammonia is ten times more toxic at 8.0 than at 7.0.

For freshwater fisheries protection it is considered that within the range pH 6.5 — 9.0 fish would not be harmed directly although the toxicity of other poisons may be affected by changes within this range. Outside this range fish may be directly affected depending on duration, species, stage of development and degree of acclimation[8].Fish are not likely to survive at less than pH 3.5 and greater than pH 11. An increasing problem of international importance is that of acid rain. Sulphur compounds are released into the atmosphere from several terrestrial sources such as volcanoes, the decomposition of sulphur-containing proteins in terrestrial and aquatic ecosystems, the burning of fossil fuels and some chemical industrial activities. SO_2 and SO_3, which results from the oxidation SO_2 in the air, dissolve in rain droplets to form acids which fall as rain, acidifying freshwaters and soils especially in areas of heavy precipitation, often far from the source of emission. It has been estimated that about 70 percent of the sulphur emitted in Britain leaves the shores of Britain during periods of dry weather with a westerly wind. This problem is receiving much attention in Scandanavia which receives airborne pollution from the industrial nations of Europe to the south.

Toxicity. Acute toxicity is probably the most popular concept of pollution; a fish-kill resulting from a poisonous spill often hits the local news headlines. In other rivers such events never occur because these are fishless due to chronic toxicity— a form of pollution which receives less public attention. The effects of toxicity — the direct poisoning of organisms — are relatively simple compared with the complexity of organic pollution and eutrophication discussed in the next section. The different physiological processes involved however are infinitely more complex. The overall ecological consequences of toxic discharges on the aquatic environment are the reduction in both the number of species and in the total number of individuals, resulting in reduced variety and abundance. Species are eliminated according to their specific tolerance to the pollutant. Different species exhibit different degrees of tolerance to different pollutants; Leeches are more tolerant of lead than of zinc whereas some species of stoneflies, a group most sensitive to organic pollution, are highly tolerant of zinc.[16] Several species of leech show an unusually high tolerance to DDT and it was found that, in some species at least, this was due to their metabolic ability to detoxify the insecticide by de-hydrochlorinating the DDT to the non-toxic DDE[17]. As the poison becomes diluted or otherwise reduced downstream of the discharge there is a successive re-appearance of species according to their degree of tolerance to the poison. In some cases the selective elimination of the less tolerant species and the resultant reduction in inter-specific competition or in predation may result in an increase in population of the more tolerant ones. Toxicity at a level which does not eliminate all life may affect the different trophic levels differentially resulting in imbalanced communities.

Toxic substances commonly polluting waters are -

Heavy metals - Pb, Ni, Cd, Zn, Cu, and Hg.

Natural organics - Phenols, Formaldehyde.

Synthetic organics - Pesticides, Herbicides, Detergents.

Inorganics - Ammonia, cyanides, fluorides, sulphides, sulphites, nitrites.

These are derived from industrial, agricultural and domestic sources.

Toxic effects may be -

Lethal - causing death by direct poisoning.

Sub-lethal - below the level which directly causes death but which may affect growth, reproduction or activity so that the population may ultimately be affected.

Acute - causing an effect (usually death) wihin a short period.

Chronic - causing an effect (lethal or sub-lethal) over a prolonged time period.

Accumulative - effect increasing by successive doses.

Much of the work on toxicity has been carried out to determine the short-term tolerance of different species, mostly adult fish, to different toxicants. The most popular and convenient way of expressing the acute toxicity of a substance is the LC50, which is that concentration of the substance which kills half the test animals in a specified period e.g. 24h, 48h or 96h. Such tests which determine the lowest concentration of a poison which kill fish in a relatively short period, 1 — 4 days, are useful in predicting the effects of acute toxicity as produced, for example, by a plug of toxic substance passing down a river. For many poisons commonly associated with industrial sewage effluents it has been found that the joint toxicity of two or more of them is simply additive [18]. By expressing the concentration of each poison present as a proportion of its LC50 value the combined toxicity can be predicted by addition; if greater than unity the water is likely to be lethal [19]. Using this technique the predicted toxicity of 24 sewage effluents on the basis of the concentrations of cyanide, ammonia, copper, zinc and phenol were compared with the toxicities as determined experimentally. In 13 out of the 18 toxic effluents, the toxicity was predicted within ±30 percent of the observed values; 6 effluents were correctly predicted to be non-toxic (acutely) and in only 2 cases was the effluent more toxic than predicted. Similarly the toxicities of fishless rivers in the industrial Midlands was largely accounted for by the additive toxicities of ammonia,copper, zinc and phenol[20].

Although these advances make possible the prediction of acute toxicity where the nature and concentration of the poisons are known, the prediction of chronic or sub-lethal effects is more difficult. Although much is now known about the rates at which fish die in toxic solutions and something about the limiting concentrations which kill fish in a relatively short period,less is known regarding the ecologically significant concentrations of poisons which can be permitted which will not adversely affect a natural population of fish continuously exposed to them. Some workers have derived permissible concentrations by applying a factor to the LC50 values.A factor of 0.1 is often quoted for metals[21],and 0.01 for pesticides[22]. Such factors are no more than intelligent guesses. A more scientifically based value may be derived by plotting the mean periods of survival at a range of concentrations. For many toxicants a curvilinear relationship is exhibited (Fig.3). In such cases the concentration at which the toxicity curve is approx. asymptotic to the time axis the incipient LC50 value may be taken as that at which the fish are likely to survive long periods. Sub-lethal effects may still occur however and some poisons are known to exert delayed effects e.g. pesticides and cadmium.

Toxicants in the aquatic environment, besides affecting the aquatic communities, pollute the water as a resource for public supply. The response of fish to pollutants, such as movement and respiratory activity, is made use of in fish alarm systems to protect water supplies[23].

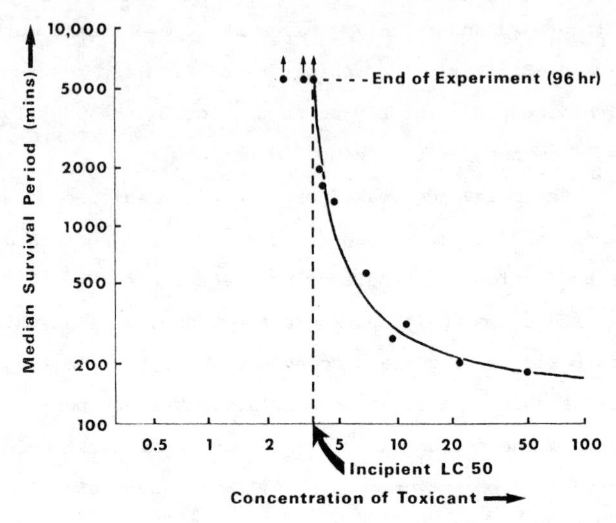

Figure 3
A Typical Toxicity Curve Showing Survival Period v Concentration

A public health hazard may arise when toxicants are concentrated via aquatic food chains used by Man. In Japan thousands of people suffered mercury poisoning by eating fish caught in Minamata Bay into which discharged an effluent containing mercury from a vinyl chloride plant. The disorder, which involves the progressive loss of co-ordination, vision, hearing and general intellect, is now known as Minamata Disease. Itai-Itai-Byo Disease, a form of cadmium poisoning causing bone deformities and fractures, has been reported as being caused by eating foods irrigated with waters polluted with cadmium from a mine tip.

Although the protection of aquatic communities from pollution generally safeguards the water as a resource for Man, because of the specificity of some toxicants care is needed in interpreting ecological changes in terms of water quality for Man's various uses. For example the presence of low concentrations of the herbicide TBA (trichlorobenzoic acid) in a river water had only slight effects on the benthic invertebrate fauna which were being used to monitor the river water quality. However water abstracted from the river was probably the cause of the failure of the tomato crop which was watered with it[24]. A marked herbicidal effect had been noted downstream of the effluent — the elimination of water-weeds — but the significance of this observation in relation to the use of the water to irrigate tomato plants, which are specifically sensitive to this herbicide at concentrations as low as 3 p.p.b., was not realized.

Approximately one third or more of the raw water abstracted for public supply in England and Wales is derived from lowland rivers many of which contain used-waters in the form of effluents from domestic and industrial sources. It is likely that on economic grounds any increase in water demand will be met from these lowland "re-use" rivers. In reviewing the public health aspects of re-using water for public supply it was considered that treatment practices and appropriate monitoring would ensure the absence of pathogenic organisms and harmful inorganic chemicals. The chronic long-term effects of the presence of thousands of organic compounds was considered difficult to predict from our present knowledge[25].

De-oxygenation . Apart from the de-oxygenation resulting from organic pollution, outlined in the next section, depletion of dissolved oxygen may result from the discharge of industrial wastes containing reducing agents such as sulphites in pulp-mill effluents and ferrous salts in mine waters. Again such effluents are associated with other pollutants which make it difficult to determine the effects of de-oxygenation itself. However dissolved oxygen is a major factor determining the distribution of invertebrates and fish in freshwaters. Species exhibit a range of tolerance to oxygen concentrations. Whereas most species are affected by the lower limits, some invertebrates such as *Chironomus riparius* have been found to survive longer at lower oxygen levels and to be adversely affected in well aerated waters [26].

It is possible to establish by laboratory tests the lowest oxygen concentration at which a species can survive for a given period of time — the incipient lethal level. This level is increased markedly at higher temperatures and higher carbon dioxide concentrations. To permit activity however to allow the species to feed, escape predation and reproduce to maintain a population, higher oxygen concentrations are needed. Furthermore some organisms such as fish are able to select the more favourable environments and avoid or leave areas of adverse conditions. Ultimately therefore the critical oxygen concentration may be that which allows a species population to remain in a given water.

In natural bodies of water, especially productive lentic waters and lowland rivers, there is, due to daytime photosynthesis, a marked diel pattern in oxygen concentrations. This diel pattern may be more significant ecologically than any individual level or the daily mean. In experiments [27] it was found that wide diel fluctuations in oxygen impaired the growth of fish which were fed an unlimited diet of earthworms. The distribution and presence or absence of *Gammarus pulex* (freshwater shrimp) in an organically polluted stream were accounted for not by the maximum, the minimum or even the mean dissolved oxygen but the number of hours during the night that the oxygen level fell below a critical value [28].

Because of the widely differing patterns of diel dissolved oxygen fluctuations, it is difficult to present dissolved oxygen criteria even for fish, on which there is much experimental data available. Tentative criteria suggested by E.I.F.A.C.[8] are expressed as percentile distributions. For a relatively tolerant species such as roach the annual 50- and 5- percentile dissolved oxygen values should be greater than 5mg l^{-1} and 2mg l^{-1} respectively. For salmonids the corresponding percentiles should be 9mg l^{-1} and 5mg l^{-1}. When other adverse conditions prevail, such as high temperatures or toxicity, higher oxygen levels would be needed.

Biotic Effects

Under this heading we shall consider the effects of discharging biogenic materials which change the nutrient status of the water so that the balance of species is substantially changed. The discharge of biodegradable organic matter such as sewage or wastes from the processing of biological materials, e.g. food, milk, textiles or paper, usually results in changes in populations detrimental to Man's interest and is therefore referred to as organic pollution. Ecologically the increased organic content results in an increase in the heterotrophic (decomposer) component of the ecosystem (Fig. 4). This causes secondary effects in the dependent food chains and by changing the autecological factors such as dissolved oxygen concentration. An increase in the concentration of plant nutrients, especially nitrates and phosphates, encourages the autotrophic (producer) component of the ecosystem (Fig.4), thus bringing about an imbalance in the community referred to as eutrophication. When eutrophication adversely affects the water quality or causes nuisance, it constitutes pollution. In other circumstances eutrophication may enhance fish production in which case it may be beneficial. Eutrophication , in this sense, is not therefore synonymous with pollution.

Organic Pollution . The discharge of putrescible organic matter to a stream causes an increase in the heterotrophic microbial population whose respiratory demand in oxidising the waste may cause a depletion of the oxygen in the water. Depending on the degree of oxygen depletion, different species are eliminated according to their tolerance, as previously outlined. In the riffle zones the natural eroding substratum may be overgrown by heterotrophic microbial growths — sewage fungus — which readily collect suspended solids causing siltation. These combined effects of the changed nutrient status, oxygen depletion and siltation affect the benthic invertebrates normally present in riffles. As a result the natural benthic community is replaced by one more typical of the decomposer community found naturally in the depositing substratum of the pools. This replacement community brings about the progressive stabilization of the organic matter, a process known as self-purification. With different stages of recovery, different benthic communities become established associated with improving conditions progressively downstream. These communities, illustrated in Fig 5, are the basis of several water pollution indicator systems e.g. Trent Biotic Index, Chandler Score, B.M.W.P. Score. Organic pollution may then be regarded as a natural response to a changed nutrient condition and as such is a good example of the homeostatic properties of ecosystems. A river which is capable of reacting in this way may be regarded as being ecologically healthy, although some of the consequences may be undesirable to Man.

In practical terms de-oxygenation is probably the most serious aspect of organic pollution in its effect on fisheries. Other nuisances may however be caused such as the growth of sewage fungus, or the infestations of midges. Of more public health significance is the increase in the population of the snail vectors of schistosomes — the blood flukes — in the warmer parts of the world.

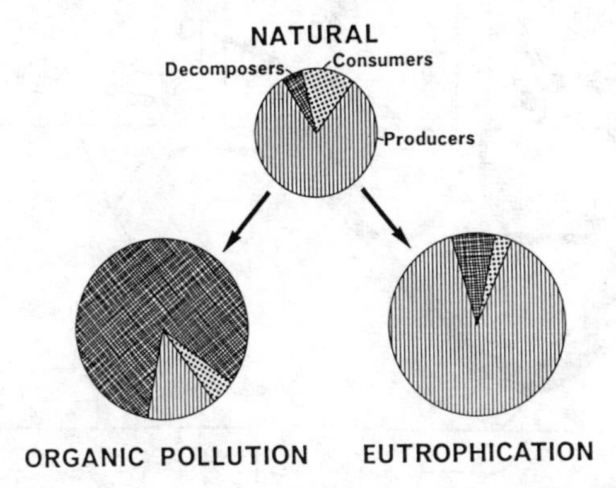

Figure 4 Ecological Imbalances Created by Organic Pollution and Eutrophication.

Eutrophication. Whereas organic pollution is usually associated with rivers, eutrophication is more commonly a phenomenon of lakes. The process of eutrophication is a natural one occurring in all lakes over a prolonged period of time — thousands of years — due to the progressive increase in fertility. This natural process may be increased by Man's activities when it is then referred to as cultural eutrophication. Successive changes in the nutrient status are associated with changes in the biota. These are usually detrimental to the several uses of the lake and water. The increased phytoplankton produces turbidity and colour. Under summer conditions algal blooms may occur which render the water toxic to fish and cattle. Macro-algae such as Cladophora and Enteromorpha proliferate and may foul the shores creating nuisance as they decay.

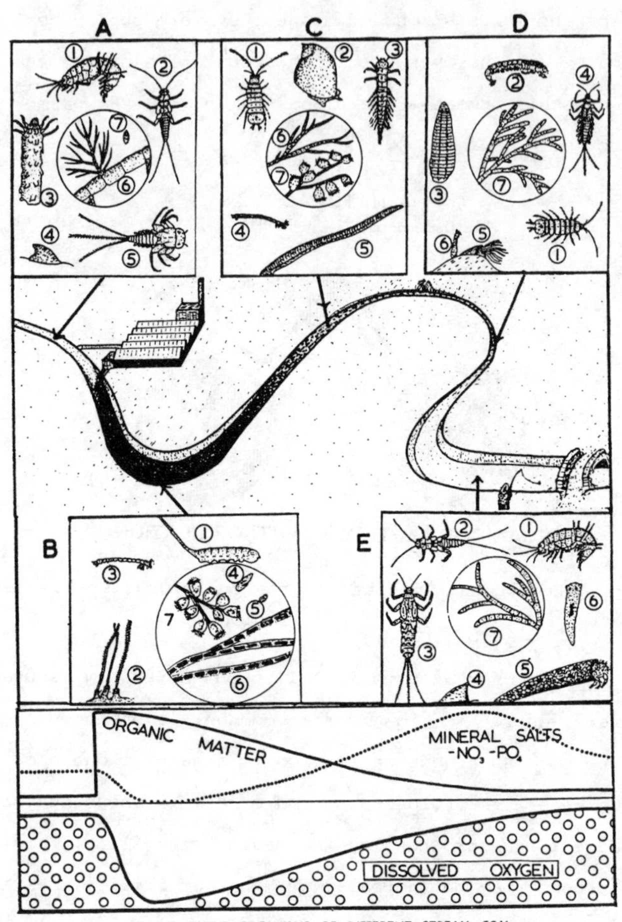

FIGURE 5

REPRESENTATIVE ORGANISMS OF DIFFERENT STREAM COMMUNITIES ASSOCIATED WITH VARIOUS DEGREES OF RECOVERING FROM ORGANIC POLLUTION

Key to organisms illustrated (micro-organisms shown within circles).

A

1. *Gammarus pulex* (Fresh-water shrimp)
2. *Nemoura* (Stone-fly nymph)
3. *Limnophilid caddis*
4. *Ancylus fluviatilis* (Limpet)
5. *Ecdyonurus* (May-fly larva)
6. *Draparnaldia* (Green alga)
7. *Cocconeis* (Diatom)

B

1. *Eristalis tenax* (Rat-tailed maggot)
2. *Tubifex* (Sludge worm)
3. *Chironomus riparius* (Blood worm)
4. *Paramecium caudatum*
5. *Colpidium*
6. *Sphaerotilus natans* (Sewage fungus)
7. *Carchesium* (Sewage fungus)

C

1. *Asellus aquaticus* (Water Hog-louse)
2. *Lymnaea pereger.* (Wandering snail)
3. *Sialis lutaria* (Alder-fly larva)
4. *Chironomus riparius* (Blood worm)
5. *Erpobdella* (Leech)
6. *Stigeoclonium* (Green alga)
7. *Carchesium* (Sewage fungus)

D

1. *Asellus aquaticus* (Water Hog-louse)
2. *Hydropsyche* (Case-less caddis larva)
3. *Glossiphonia* (Leech)
4. *Baetis rhodani* (May-fly nymph)

5 & 6. *Simulium ornatum* Pupa & larva of Buffalo gnat

7. *Cladophora* (Blanket weed)

E

1. *Gammarus pulex* (Fresh-water shrimp)
2. *Nemoura* (Stone-fly nymph).
3. *Ephemerella* (May-fly nymph)
4. *Ancylus fluviatilis* (Limpet)
5. *Stenophylax* (Caddis larva)
6. *Dugesia* (Flat-worm)
7. *Cladophora* (Blanket weed)

(Reproduced by permission of Effluent and Water Treatment Journal)

The eutrophication of lakes and reservoirs used for public supply renders the water less suitable because of the increase in bacterial numbers and the possibility of tastes and odours produced by some phytoplankton. The removal of phytoplankton increases the cost of treatment at the water-works. In the rapidly flowing rivers phytoplankton do not cause problems but under nutrient-enriched conditions attached filamentous algae such as Cladophora may develop profuse growths to cause a nuisance. In some impounded brackish fenland streams blooms of the phytoflagellate Prymnesium parvum occur which cause fish mortality due to the secretion of toxins.

Of the macronutrients involved in algal growth (C, N and P) phosphorus is probably more often the limiting one in natural waters. It has however proved difficult to establish by culture methods the critical concentration to which P needs to be reduced to prevent eutrophic conditions occurring. In culture the growth-limiting concentration for some algae was found to be as low as 0.008 mg P l^{-1}. There is some evidence that in lakes P is made available for synthesis by the lysis of dead algal cells as they sink through the euphotic zone. Some algae assimilate P in excess of their requirements when it is available and this is then inherited by their progeny at cell division. In consequence dense phytoplankton populations can be maintained in waters in which P is not detectable. On the basis of data obtained during a two-year study of 17 lakes in the U.S.A.[29], it was predicted that algal blooms of nuisance level could occur in lakes having, at the time of the spring turn-over, an inorganic N concentration exceeding 0.3 mg N l^{-1} and an inorganic P concentration exceeding 0.015 mg P l^{-1}.

It was considered that eutrophication was irreversible. More recent experience however has shown that this is not always so. The diversion of the discharges to Lake Washington, which had caused serious eutrophication, resulted in a marked improvement in conditions[30]. However in lakes in which bottom sludge deposits have accumulated, which can act as a nutrient reservoir, the diversion of the nutrient input alone may not have immediate results; this was the case in Lake Trummen (Sweden) where only after dredging the bottom sediments was there any improvement[31].

References

1 A.S.Wisdom, The Law on the Pollution of Waters, Shaw, London (1956).

2 H.A.Hawkes, Biological Aspects of River Pollution. In - L.Klein (Ed.) River Pollution,2 Causes and Effects, Butterworths,London (1962) Chapter8,p399.

3 M.W.Holdgate, A Perspective of Environmental Pollution, Cambridge University Press, Cambridge (1979).

4 L.Klein, River Pollution, 2 Causes and Effects, Butterworths (1962) Chapter 3 p.25.

5 F.T.K. Pentelow, Rep. Salm. Freshw. Fish. 1949,London.

6 D.W.M. Herbert and J.C. Merkens, Int. J. Air Wat. Pollut., 1961, 5, 46.

7 Ministry of Technology, Water Pollution Research, 1961, H.M.S.O., London (1962).

8 J.S. Alabaster and R. Lloyd, Water Quality Criteria for Freshwater Fish,Butterworths,London (1980).

9 C.M Tarzwell, Water Quality Criteria for Aquatic Life,

In Biological Problems in Water Pollution, U.S. Dept. of Health, Educ. and Welfare, Public Health Service, R.A. Taft San. Eng. Cent., Cincinnati,Ohio (1957) 248.

10 T.E. Langford, Hydrobiologia,1975, 47, 91.

11 K. Patalas, Summary Reports of 7th Hydrobiologists' Congress, Swinoujscie, Sept. 1967.

12 V.M. Brown, D.H.M. Jordan and B.A. Tiller, Wat. Res. 1967, 1, 587.

13 World Health Organization, Sodium, Chlorides and Conductivity in Drinking Water. W.H.O., Copenhagen (1979).

14 A.D. Harrison, Verh. int. Ver. Limnol.,1958, 13, 603.

15 J.W. Hargreaves, E.J.K. Lloyd and B.A. Whitton, Freshwater Biol., 1975, 5, 563.

16 K.E. Carpenter, Ann. appl. Biol., 1924, 11, 1.

17 R.T. Sawyer, Leeches (Annelida, Hirudinea) In Pollution Ecology of Freshwater Invertebrates, Eds. C.W. Hart and S.L.H. Fuller,Academic Press,London (1947)

18 R. Lloyd, Ann. appl. Biol.,1961, 49, 535.

19 V.M. Brown, Wat. Res., 1968, 2, 723.

20 D.W.M. Herbert, D.H.M. Jordan and R. Lloyd, J. Proc. Inst. Sew. Purif., 1965,(6), 569.

21 J.R.E. Jones, Fish and River Pollution, Butterworths, London (1964).

22 A.V. Holden, J. Proc. Inst. Sew. Purif., 1964,361.

23 W.S.G. Morgan, Prog. Wat.Tech.,1978, 10, 395.

24 H.A. Hawkes, River-bed Animals Tell-tales of Pollution, In Biosurveillance of River water Quality, Eds. H.A. Hawkes and J.G. Hughes,Proceedings of Section K of the British Association for the Advancement of Science, Aston,1977, 66.

25 Central Water Planning Unit,The Significance of Synthetic Chemicals in Rivers used as a Source of Drinking Water. Tech. Note 14. Reading (1976) .

26 H.M.Fox and A.E.R.Taylor, Proc. roy. Soc. B, 1955, 143, 214.

27 N.E.Stewart, The influence of oxygen concentration on the growth of juvenile large mouth bass. M.Sc. Thesis, Oregon State University Library,Corvallis,Oregon (1962). Quoted in P. Doudorof and C.E. Warren,Dissolved Oxygen Requirements of fishes. In Biological Problems in Water Pollution. U.S. Dept. of Health,Educ. and Welfare. Public Health Service, R.A. Taft San. Eng. Cent., Cincinnati, Ohio(1965), 145.

28 H.A. Hawkes and L.J. Davies, In E.A. Duffey and A.S. Watt (Eds.) The scientific management of animal and plant communities for conservation,Blackwell,Oxford (1971) 271.

29 C.N. Sawyer, J.Wat.Pollut.Control Fed., 1966, 38,737.

30 W.T. Ednonson, Verh. internat. Verein. Limnol. 1972,18,284.

31 C. Gelin, The restoration of freshwater ecosystems in Sweden. In M.W. Holdgate and M.J. Woodman,(Eds.) The breakdown and restoration of ecosystems,Plenum, New York (1978) 332.

8
Important Air Pollutants and Their Chemical Analysis

By R. M. Harrison
DEPARTMENT OF ENVIRONMENTAL SCIENCES, UNIVERSITY OF LANCASTER, LANCASTER LAI 4YQ, U.K.

1. INTRODUCTION

Before any detailed consideration of this topic, some description of basic terminology is worthwhile. Air pollutants may exist in gaseous or particulate form. The former include substances such as sulphur dioxide and ozone, and their concentrations are expressed most commonly either in mass per unit volume ($\mu g\ m^{-3}$ of air), or as a volume mixing ratio (1 ppm = 10^{-6}; 1 ppb = 10^{-9}). Particulate air pollutants are very diverse in character, including both organic and inorganic substances with diameters ranging from <0.01 — >100 μm. Since very fine aerosol particles grow rapidly by coagulation, and large particles sediment rapidly under gravitational influence, the major part exists in the 0.1 — 15 μm range. Particles <*ca*. 2 μm are generally formed by growth of smaller particles generated by condensation processes, whilst larger particles arise from mechanical disintegration processes. Thus the major part of the aerosol <2 μm comprises man-made components (e.g. lead from motor exhausts, ammonium sulphate from atmospheric oxidation of sulphur dioxide), whilst the >2 μm material is frequently natural in origin (e.g. wind-blown soil, marine aerosol). This is obviously not a rigid division. The size distribution typical of ambient aerosols is shown in Figure 1.

Pollutants are emitted from sources, whilst they are removed from the atmosphere by sinks. The major sink processes for air pollutants are chemical reactions and wet and dry deposition processes. Thus a sink for sulphur dioxide is atmospheric oxidation to sulphuric acid. The sulphuric acid, possibly after neutralisation by gaseous ammonia, may be removed from the air by incorporation in rainwater, either within the cloud layer (known as rainout), or by falling raindrops (known as washout), or by dry deposition onto surfaces by gravitational settling, impaction or Brownian diffusion, dependent upon particle size.

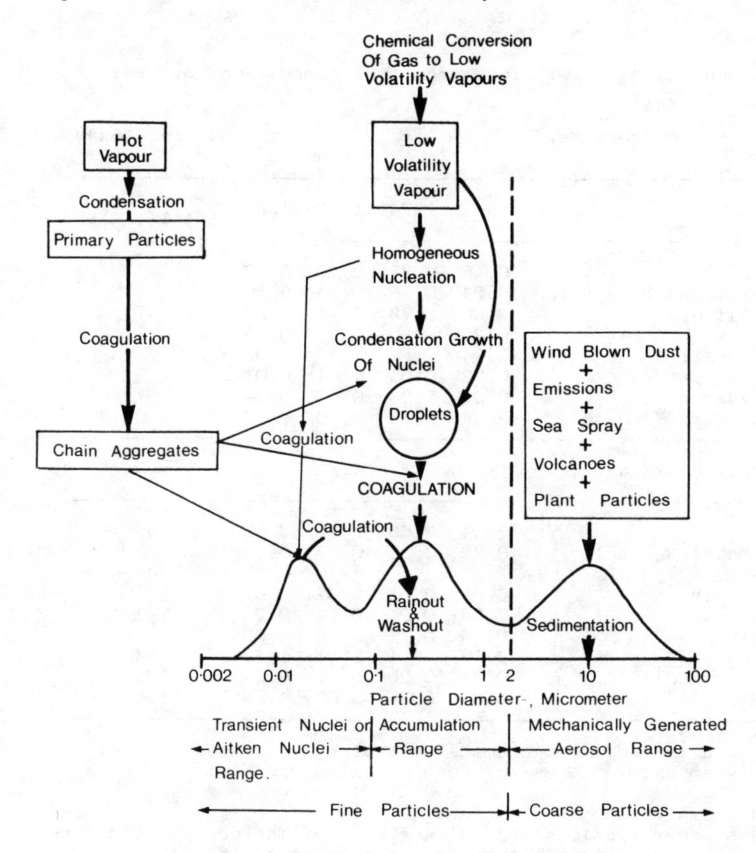

Figure 1 Schematic diagram of the size distribution (expressed as surface area per increment in particle diameter) and formation mechanisms for atmospheric aerosols.*

2. SPECIFIC AIR POLLUTANTS

2.1 Sulphur Dioxide

The major source of sulphur dioxide is the combustion of fossil fuels containing sulphur. These are predominantly coal and fuel oil, since natural gas, petrol and diesel fuels have a relatively low sulphur content. Table 1 shows an emission inventory for sulphur dioxide, both by type of consumer and by type of fuel. The marked decline in domestic emissions during the 1970's arose primarily from the switch by many consumers to natural gas during this period.

*Reproduced from Atmos. Environ., 12, 136 by permission of Pergamon Press Ltd.

Table 1

U.K. Sulphur dioxide emissions from fuel combustion: by type of consumer and fuel[1]

(a) By type of consumer — Million tonnes

	1971	1974	1977	1980	Percentage of total in 1980
Domestic	0.46	0.35	0.29	0.22	5
Commercial/public service[2]	0.34	0.26	0.24	0.20	4
Power stations	2.80	2.78	2.74	2.87	61
Refineries	0.25	0.30	0.27	0.28	6
Other industry[3]	1.94	1.59	1.39	1.05	22
Rail transport	0.02	0.02	0.01	0.01	< 1
Road transport	0.05	0.05	0.05	0.04	1
All consumers	5.86	5.36	5.00	4.68	100

(b) By type of fuel

	1971	1974	1977	1980	Percentage of total in 1980
Coal	3.04	2.56	2.79	3.02	65
Solid smokeless fuel	0.20	0.16	0.13	0.10	2
Petroleum					
Motor spirit	0.02	0.01	0.02	0.01	1
Derv	0.03	0.04	0.04	0.03	1
Gas oil	0.16	0.17	0.14	0.09	2
Fuel oil	2.16	2.11	1.61	1.13	24
Refinery fuel	0.25	0.30	0.27	0.28	6
All fuels	5.86	5.36	5.00	4.68	100

[1]Excludes emissions from chemical and other processes, which probably amount to a few per cent of emissions from fuel combustion.
[2]Includes miscellaneous consumers.
[3]Excludes power stations and refineries, but includes agriculture
Source: Warren Spring Laboratory, Department of Industry.

Ambient air concentrations of sulphur dioxide measured by the "National Survey of Air Pollution" of the Warren Spring Laboratory are shown in Table 2.

Table 2

Sulphur dioxide: daily concentrations[1] at U.K. urban sites for the winter period[2]: by region and country

	Average concentration μg/m^3			
	74/75	77/78	78/79	79/80
North	68	60	57	51
Yorkshire and Humberside	100	84	94	85
East Midlands	73	75	84	75
East Anglia	65	51	62	60
Greater London	117	95	100	81
South East (excluding Greater London)	62	53	59	50

Table 2 continued

	Average concentration μg/m^3			
	74/75	77/78	78/79	79/80
South West	47	41	49	41
West Midlands	87	75	86	67
North West	96	86	91	83
England	86	73	80	69
Wales	56	53	54	47
Scotland	63	57	58	53
Northern Ireland	60	45	41	44
United Kingdom	82	70	76	65

[1]Due to instrumental and operational problems and site replacement winter averages are usually available for about 800-900 of the 1,050 or so urban sites monitored each year; since the averages are not based on precisely the same sites from winter to winter the figures are not strictly comparable between winters.
[2]From October to the following March.
Source: 'National Survey of Air Pollution' Warren Spring Laboratory, Department of Industry.

Whilst emissions of this pollutant have fallen by 20% since 1960, average urban concentrations have reduced by over 60% over the same period. This arises from a shift from low level to high level sources. The low level sources (e.g. domestic chimneys) affect the local urban area, whilst a high level source, such as a power station, disperses its pollutants over a far greater area. Whilst this trend in emission heights has benefited British urban areas, it has exacerbated another problem, that of pollutant transport. Sulphur dioxide, and other pollutants, emitted at a high level may be transported over very large distances by the atmosphere. During such transport processes, oxidation of sulphur and nitrogen oxides to sulphuric and nitric acids proceeds, hence giving rise to an "acid rain" problem at great downwind distances. This problem is most acute in Scandinavia which itself emits little sulphur dioxide. Several environmental problems are associated with the acid rain, including the killing of fish in acidified lake waters and the leaching of nutrients from soils.

The adverse effects of sulphur dioxide itself include damage to human respiratory function, especially when exposure is in conjunction with particulates as in the London smogs of the 1950's. Sulphur dioxide is also damaging to plants at modestly elevated concentrations. Whilst Britain does not set ambient air quality standards for pollutants, the U.S. Environmental Protection Agency standards for this pollutant are 80 $\mu g\ m^{-3}$ (annual average) and 365 $\mu g\ m^{-3}$ (24h average) and the World Health Organization long-term goal is 60 $\mu g\ m^{-3}$ (annual mean) with 98% of measurements $<200\ \mu g\ m^{-3}$.

2.1.1. Chemical Analysis

Sulphur dioxide may be determined by many procedures of widely differing sensitivity and specificity. The method used in the "National Survey of Air Pollution" involves absorption of the sulphur dioxide in hydrogen peroxide. The resultant acid is determined by acid-base titration or by conductivity measurement. Although acidic or basic compounds interfere, these techniques can yield useful results in urban areas where sulphur dioxide concentrations are high and levels of interfering substances are low. In rural areas this is not the case, due to natural production of ammonia, and misleading results are obtained. The West-Gaeke technique, in which sulphur dioxide is collected by reaction with potassium or sodium tetrachloromercurate (II) forming the disulphitomercurate (II), and then determined colorimetrically after addition of acidic pararosaniline methyl sulphonic acid, has been refined to the point where the effects of known interfering substances have been minimised or eliminated.

$$[HgCl_4]^{2-} + 2SO_2 + 2H_2O \longrightarrow [Hg(SO_3)_2]^{2-} + 4Cl^- + 4H^+$$

By using flow systems, continuous monitors have been built using this reaction but the response time is relatively long, of the order of several minutes, and short-lived concentration peaks are not accurately measured.

In the flame photometric sulphur analyser, gaseous sulphur compounds are burned in a reducing hydrogen-air flame and the emission of the S_2 species at 394 nm is measured. The technique is sensitive (ppb levels are measurable), specific to sulphur and has a very fast response as polluted air may be passed directly to the flame. The sensitivity is similar for all volatile sulphur compounds and in urban air the level of total volatile sulphur, as given by a flame photometric analyser, approximates closely to the sulphur dioxide level. When several sulphur compounds are present in air at comparable concentrations, gas chromatographic separation is possible from a small air sample at ppb levels using the flame photometric analyser as detector.

An alternative technique is specific for measurement of sulphur dioxide at levels down to 1 ppb. Fluorescence, excited by radiation in the region of 214 nm, is measured (Fig. 2) and a very wide linear range of response is found. Instruments covering the range of 1-5000 ppm and 0-0.5 ppm of SO_2 are available.

The commonly used techniques for analysis of SO_2 and other pollutants are summarised in Table 3.

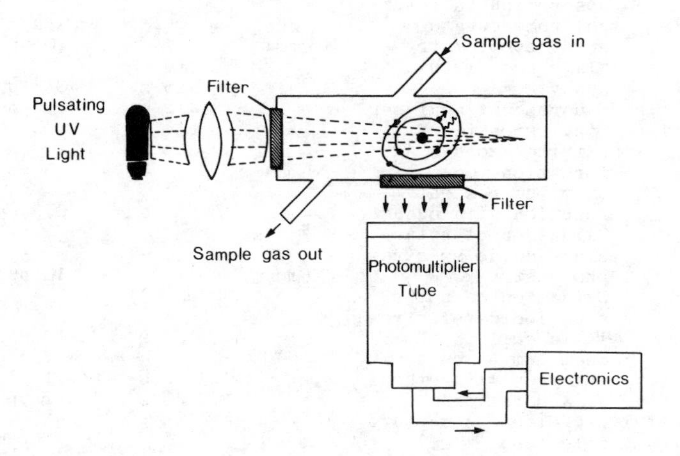

Figure 2. Pulsed fluorescent analyser for sulphur dioxide.

Table 3. Summary of commonly employed methods for measurement of gaseous air pollutants.

Pollutant	Measurement technique	Sample collection period	Response time[a] (continuous techniques)	Minimum detectable concentrations
Total hydrocarbons	Ndir		5 s	1 ppm (as hexane)
	Flame ionisation analyser		0.5 s	10 ppb (as methane)
Specific hydrocarbons	Glc	b		1 ppb
Carbon monoxide	Ndir		5 s	0.5 ppm
	Catalytic methanation /fid	c		10 ppb
	Electrochemical cell		25 s	1 ppm
Sulphur dioxide	Absorption in hydrogen peroxide/titration	24 h		2 ppb

Table 3 continued

Pollutant	Measurement technique	Sample collection period	Response time (continuous techniques)	Minimum detectable concen-trations
	Absorption in hydrogen peroxide/conductivity determination		3 min	10 ppb
	Absorption in tetrachloromercurate/spectrophotometry	15 min		10 ppb
	Flame photometric analyser		25 s	0.5 ppb
	Fluorescent analyser		2 min	0.5 ppb
Oxides of nitrogen	Conversion to nitrite/azo dye formation	30 min		5 ppb
	Chemiluminescent reaction with ozone		1 s	0.5 ppb
Ozone	Oxidation of potassium iodide/spectrophotometry	30 min		10 ppb
	Oxidation of potassium iodide/electrolytic cell		30 s[d]	10 ppb
	Chemiluminescent reaction with ethene		3 s	1 ppb
	Uv absorption		30 s	3 ppb
Peroxyacetyl nitrate	Glc/electron capture detection	c		1 ppb

(a) Time taken for a 90 per cent response to an instantaneous concentration change
(b) Grab samples of air collected in an inert container and concentrated prior to analysis
(c) Instantaneous concentrations measured on a cyclic basis by flushing the contents of a sample loop into the instrument
(d) Time for a 75 per cent response

2.2 Oxides of Nitrogen

The most abundant nitrogen oxide in the atmosphere is nitrous oxide, N_2O. This is chemically rather unreactive and is formed by natural microbiological processes in the soil. It is not normally considered as a pollutant, although it does have an effect upon stratospheric ozone concentrations and there is concern that use of nitrogenous fertilisers may be increasing atmospheric levels of nitrous oxide.

The pollutant nitrogen oxides of concern are nitric oxide, NO, and nitrogen dioxide, NO_2. By far the major proportion of emitted NO_x, as the sum of the two compounds is known, is in the form of NO, although most of the atmospheric burden is usually in the form of NO_2. The major conversion mechanism is the very rapid reaction of NO with ambient ozone, the alternative third order reaction with molecular oxygen being relatively very slow at ambient air concentrations.

The major source of NO_x is the high temperature combination of atmospheric nitrogen and oxygen in combustion processes, there being also a lesser contribution from combustion of nitrogen contained in the fuel. An emission inventory for the U.K. appears in Table 4.

Table 4. Nitrogen oxides: estimated U.K. emissions by source

(Thousand tonnes (as NO_2))

	1971	1974	1977	1980	Percentage of total in 1980
Domestic	49	50	47	48	3
Commercial and industrial	690	624	502	422	23
Power stations	755	729	786	849	46
Incineration and agricultural burning	6	6	8	8	< 1
Road vehicles					
petrol engined	247	272	286	316	17
diesel engined	156	166	171	176	9
Railways	54	49	43	44	2
All emissions	1,956	1,897	1,843	1,862	100

Typical ambient air concentrations of NO_x are normally within the range 5-100 ppb (roughly 10-200 $\mu g\ m^{-3}$) in urban areas and <20 ppb at rural sites. The U.S. Environmental Protection Agency (U.S.E.P.A.) ambient air quality standard for NO_2 is 100 $\mu g\ m^{-3}$ (annual average). The direct effects of exposure to oxides of nitrogen include human respiratory tract irritation and damage to plants. Indirect effects arise from the essential role of NO_2 in photochemical smog reactions, and its oxidation to nitric acid contributing to acid rain problems.

2.2.1 Chemical Analysis

In the analysis of oxides of nitrogen the common wet chemical techniques are based upon the method of Saltzman, which is reported to be unreliable, particularly at low levels. Nitrogen dioxide is collected in a fritted bubbler containing an aqueous mixture of sulphanilic acid (4-aminobenzene sulphonic acid), N-(1-naphthyl)ethylenediamine dihydrochloride and acetic acid. The first stage is conversion of NO_2 to nitrite, the anticipated stoichiometry being given by

$$2NO_2 + H_2O \longrightarrow HNO_2 + HNO_3$$

Diazotisation of the sulphanilic acid occurs, leading to the formation of an azo dye which is determined spectrophotometrically. Calibration of the technique using nitrite was found to be inconsistent with that using gaseous NO_2. Instead of the expected equivalence of 0.5 mol of nitrite to 1 mol of NO_2, Saltzman found values in the range 0.48-0.72 mol of nitrite and the upper figure of 0.72 is generally accepted. Saltzman postulated a concurrent reaction of sulphanilic acid directly with NO_2 to give a diazo product via N-nitroso-sulphanilic acid. Other drawbacks include the highly variable efficiency of different bubblers for collection of NO_2, and interferences from ozone and sulphur dioxide. To measure NO_x (NO + NO_2) prior oxidation of NO is necessary, normally with acidic potassium permanganate. This process is not quantitative and an efficiency of 70 per cent has been suggested.

Problems have also arisen with a related procedure for NO_2 determination, at one time used in the US as a reference method for NO_2 in ambient air. In a modification of the method of Jacobs and Hochheiser, nitrogen dioxide is collected by conversion to stable sodium nitrite in a bubbler containing 0.1 M aqueous sodium hydroxide over a 24h sampling period. Subsequent diazotisation of sulphanilamide (4-aminobenzene sulphonamide) and coupling with N-(1-napththyl)ethylenediamine (N-(naphthalene-1 ethane-1,2-diamine) in an acid medium leads to formation of an azo dye which is determined spectrophotometrically at 540 nm. A collection efficiency of 35 per cent was assumed, and results corrected appropriately, the method being calibrated with standard nitrite solution. The advent of permeation tubes, however, allowed dynamic calibration with gaseous NO_2, and collection efficiencies from 15-70 per cent were found, dependent upon the NO_2 concentration. An interference from nitrogen oxide was also found and doubts regarding the method caused a delay in implementing regulations controlling NO_x emission from stationary sources.

The currently favoured technique for determination of oxides of nitrogen is based upon the chemiluminescent reaction of nitrogen oxide and ozone to give an electronically excited nitrogen dioxide which emits light in the 600-3000 nm region with a maximum intensity near 1200 nm:

$$NO + O_3 \longrightarrow NO_2^* + O_2$$

$$NO_2^* \longrightarrow NO_2 + h\nu$$

In the presence of excess ozone, the light emission varies linearly with the concentration of nitrogen oxide from 1 ppb to 10^4 ppm. Theapparatus is shown in Fig. 3.

The method is believed free of interference for measurement of NO and may be used to measure NO_x by prior conversion of NO_2 to NO in a tube heated to about 650°C. Oxidation of ammonia to nitrogen oxide in the tube is possible but higher temperatures are generally required and some other nitrogen-containing compounds may similarly interfere but this is rarely significant in ambient monitoring.

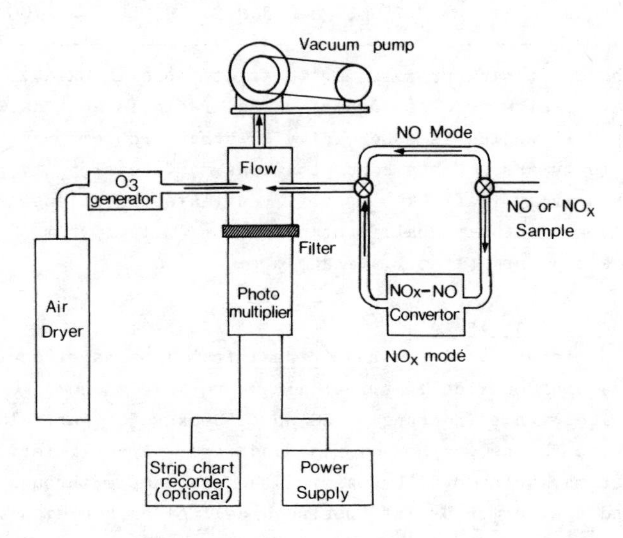

Figure 3. Chemiluminescent analyser for oxides of nitrogen.

2.3 Carbon Monoxide

This pollutant is very much associated with petrol-engined vehicles (see Table 5). Car exhaust gases contain several percent carbon monoxide under normal running conditions, and even greater concentrations when cold and choked. Most combustion processes are relatively efficient and cause rather little CO emission.

Table 5 Carbon monoxide : estimated U.K. emissions by source (thousand tonnes)

	1971	1974	1977	1980	Percentage of total in 1980
Domestic	1,016	840	657	528	6
Commercial and industrial	94	89	75	59	1
Power stations	46	45	47	49	1
Incineration and agricultural burning	214	214	223	224	3
Road vehicles					
petrol engined	6,007	6,617	6,959	7,689	87
diesel engined	213	226	234	241	3
Railways	19	18	16	16	< 1
All emissions	7,608	8,048	8,211	8,805	100

The major sink process for CO is atmospheric oxidation to CO_2, a rather slow process involving free radicals such as OH and HO_2. Carbon monoxide is essentially an urban problem due to its toxicity to humans. Urban concentrations may occasionally reach 50 ppm in heavily trafficked, poorly ventilated locations. The U.S.E.P.A. ambient air quality standard for CO is 10 ppm (8 hour average) and 40 ppm (1 hour average).

2.3.1 Chemical Analysis

Non-dispersive infra-red (*vide infra*) is sensitive enough to measure carbon monoxide in street air where levels encountered normally lie within the range 1-50 ppm. Because of partial overlap of absorption bands, carbon dioxide and water vapour interfere. The latter may be removed by passing the air sample through a drying agent, and the former by interposing a cell of carbon dioxide between the sample and reference cells and the detectors. A gas chromatographic technique, although not allowing continuous measurement of CO, has a substantially higher sensitivity. The basis of the method is catalytic reduction of the carbon monoxide to methane and detection by fid (Fig. 4). This principle is used in commercial instruments which measure methane, carbon monoxide and total hydrocarbons.

Hydrogen carrier gas flushes a small air sample from a sample loop through a stripper column, (C1) packed with an adsorbent porous polymer, for sufficient time to allow passage of methane and carbon monoxide but not the heavier compounds which are subsequently removed by back-flushing. Passage of the methane and carbon monoxide through a chromatographic column containing molecular sieves

(C3) causes separation of these compounds which are passed to the fid via a catalytic methanator. Sub-ppm levels are readily determined. Modification of the instrument allows passage of C_2 hydrocarbons through the stripper column, and these are separated on a second column of porous polymer and passed directly to the detector. Total hydrocarbons are determined by passage of a 10 cm^3 air sample directly into the fid (valve B). The carrier gas is air, purified by catalytic oxidation of impurities and this, rather than more conventional carrier gases, eliminates problems resulting from radical changes in flame characteristics upon introduction of the polluted air sample into the detector.

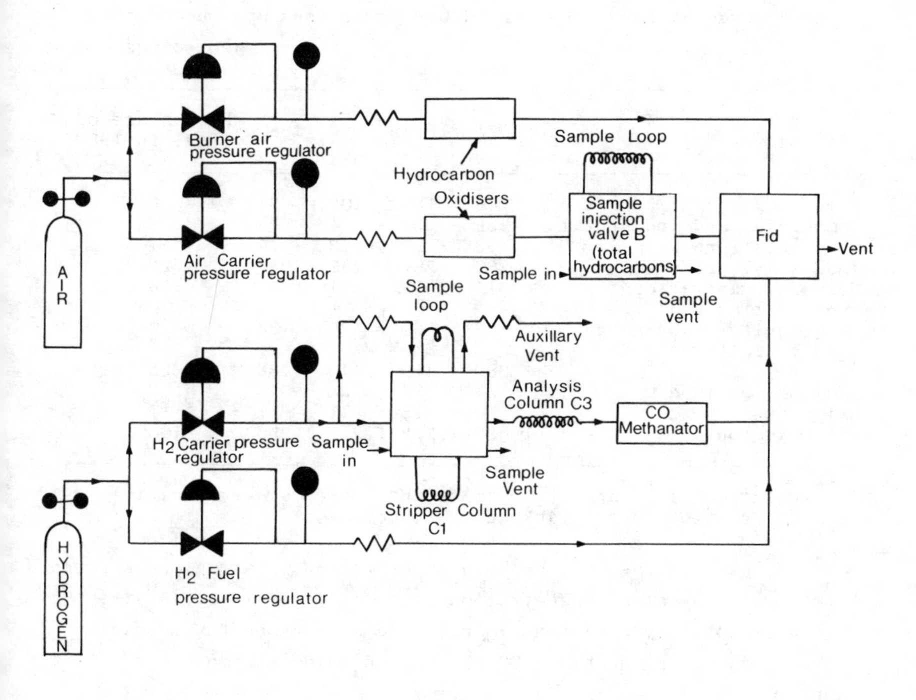

Figure 4. Analyser giving sequential measurements of methane, carbon monoxide and total hydrocarbons.

An analyser for continuous determination of carbon monoxide at levels down to 1 ppm uses an electrochemical cell. Gas diffuses through a semi-permeable membrane into the cell and at an electrode CO is oxidised to CO_2 at a rate proportional to the concentration of CO in the air. The response time is fairly short and interferences from other air pollutants at normally encountered levels are minimal. Analysers based upon electrochemical cells are also available for measurement of sulphur dioxide and oxides of nitrogen.

2.4 Hydrocarbons

Major sources of hydrocarbons in air (Table 6) are the evaporation of solvents and fuels, and the partial combustion of fuels. Obviously such processes give rise to an enormous range of individual compounds, and careful analytical work has shown measurable levels of in excess of 200 hydrocarbon compounds in some ambient air samples.

Table 6 Hydrocarbons : estimated U.K. emissions by source (thousand tonnes)

	1971	1974	1977	1980	Percentage of total in 1980
Domestic	166	130	106	81	6
Commercial and industrial	39	37	34	28	2
Power stations	13	12	13	14	1
Industrial processes	287	287	287	287	22
Solvent evaporation	311	311	311	311	24
Incineration and agricultural burning	36	37	38	38	3
Road vehicles					
petrol engined[1]	380	419	440	486	38
diesel engined	34	36	38	39	3
Railways	13	12	11	11	1
All emissions	1,280	1,282	1,277	1,295	100

[1] Including mobile evaporative emissions, which are evaporative losses from the petrol tank and carburettor

Methane is the most abundant hydrocarbon in air at background sites, existing at a concentration of ca. 1-4 ppm in clean air. Levels of total hydrocarbons in polluted air may be far in excess of this value, approaching 100 ppm (as CH_4) in extreme urban samples.

Because of the diversity of hydrocarbons in air, it is not possible to make clear statements upon toxic levels. Some individual compounds such as benzene are of high toxicity to humans, whilst others, most specially ethene, are damaging to growing plants The major problem associated with hydrocarbons, however, is their important role in promoting photochemical smog formation, and it is for this reason that hydrocarbon emissions are subject to strict controls in many countries.

2.4.1. Chemical Analysis

Techniques for determination of total hydrocarbons (THC) in air do not require an exceptional sensitivity since a natural background of methane exists throughout the troposphere (atmosphere below ∿ 10 km) and anthropogenic release may contribute to substantially elevated local levels of hydrocarbons.

For some years non-dispersive infrared (ndir) was a technique much used for total hydrocarbon measurements. Using a long-path cell, the ir absorbance of polluted air at the wavelength corresponding to the C-H stretching vibration is continuously determined relative to that of air containing no hydrocarbons. This is achieved without dispersion of the ir radiation by using cells containing n-hexane vapour at a reduced pressure as detectors for two beams which are chopped at a frequency of 10 Hz and passed through sample and reference cells. Absorption of radiation by the hexane causes a differential pressure between the two detector cells which is sensed by a flexible diaphragm between them. The method is of barely adequate sensitivity for ambient measurements and shows a very variable sensitivity towards different hydrocarbons as a result of the varying wavelengths and intensities of absorption by different compounds. The sensitivity per carbon atom in the molecule is similar for most alkenic compounds except methane but is substantially less for alkenes and aromatics and the method is almost totally insensitive to acetylene.

The flame ionisation analyser, using the conventional flame ionisation detector (fid) for measurement of THC is far more sensitive than ndir. Sensitivity per carbon atom of individual hydrocarbons varies very little, whilst oxygenated and halogenated compounds cause a lower response. Water vapour, carbon dioxide and carbon monoxide do not significantly affect the reading.

Determination of specific hydrocarbons in ambient air normally requires a pre-concentration stage in which air is drawn through an adsorbent, such as a porous polymer or activated carbon, or a tube where freeze-out of the compounds by reduced temperature occurs followed by injection into a gas liquid chromatograph. Excellent separations of many compounds have been achieved, the best results coming from use of capillary columns. Detection may be by flame ionisation, or by passage into a mass spectrometer (gc-ms), the latter technique allowing a more positive identification of individual compounds.

2.5 Secondary Pollutants

2.5.1 Ozone

Atmospheric reactions involving oxides of nitrogen and hydrocarbons cause the formation of a wide range of secondary products. The most important of these is ozone. In severe photochemical smogs, such as occur in Southern California, levels of ozone may exceed 400 ppb.

In Britain, we do not experience the classic Los Angeles type of smog. Nonetheless, the same chemical processes give rise to elevated levels of ozone, often in a regional phenomenon extending over hundreds of miles simultaneously. Thus hydrocarbon and NO_x emissions over wide areas of Europe react in the presence of sunlight causing large scale pollution, which is further extended by atmospheric transport of the ozone. The phenomenon is crucially dependent upon meteorological conditions, and hence in Britain is observed on only perhaps 10-30 days in each year on average. Concentrations of ozone measured at ground level commonly exceed 100 ppb during such "episodes", and have on one severe occasion been observed to exceed 250 ppb in Southern England. These levels may be compared with a natural background of ozone at ground level arising from downward diffusion of stratospheric ozone of 20-50 ppb.

There is evidence of adverse health effects arising from human exposure to ozone, especially in association with other pollutants such as sulphur dioxide. The U.S.E.P.A. ambient air quality standard is 120 ppb, not to be exceeded as an hourly average more than once per year. Damage to crop plants may occur at levels below this, with some varieties showing adverse effects at levels as low as 50 ppb. Economic losses to crops in the United States due to ozone damage are considerable.

2.5.1.1 Chemical Analysis

No specific wet chemical method is available for ozone analysis and concentrations of "oxidant" in the air have been determined by measurement of the capacity of an air sample to oxidise neutral buffered potassium iodide. Thus the method is sensitive not only to ozone but to other oxidising substances in the air including peroxyacyl nitrates, and nitrogen dioxide with a lesser sensitivity. The overall reaction with ozone was assumed to be as below with a consequent stoichiometry of I_2/O_3 equal to 1.0.

$$O_3 + 2H^+ + 2I^- \rightarrow I_2 + H_2O + O_2$$

$$I_2 + KI \rightarrow KI_3$$

Some workers, however, have found a stoichiometry for I_2/O_3 of 1.5 and observed an enhanced formation of iodine and a reduced oxygen formation with increasing pH and postulated a more complex mechanism. The feasibility of iodide oxidation to iodate has been shown, and the possibility of catalysis of this reaction by a glass frit. Hence, in addition to the lack of specificity of the reagent the stoichiometry of the reaction with ozone is in doubt. The use of a borate, rather than a phosphate buffer has recently been shown to overcome many of these problems.

Automated methods have been devised to allow continuous measurement of oxidant levels using these reactions. One method is based upon continuous spectrophotometric monitoring of triiodide at 352 nm, whilst a second procedure uses an iodide specific ion electrode to monitor the disappearance of I^-. Alternatively, continuous amperometric conversion of trihalide to halide ions may be achieved using a galvanic or electrolytic cell. In the galvanic cell, a carbon anode is oxidised whilst tribromide is reduced at the cathode and the ozone is determined coulometrically. In the electrolytic cell, formation of iodine allows passage of current between polarised platinum electrodes, the current flow being a linear function of iodine concentration. Although continuous measurements may be made using these techniques, response times are long - of the order of minutes.

Appropriate chemical filters improve the selectivity of methods based upon the oxidation of neutral buffered halide. Sulphur dioxide produces a negative interference equal to the equivalent molar concentration of oxidant and may be eliminated by incorporating a chromic acid paper absorber in the sampling line. The chromic acid does, however, oxidise nitrogen oxide to nitrogen dioxide causing further interference in the presence of NO.

Ozone may be continuously and specifically determined by measurement of the light emitted by the chemiluminescent reaction of ozone and ethene:

$$C_2H_4 + O_3 \longrightarrow \text{Ozonide} \longrightarrow CH_2O^*$$

$$CH_2O^* \longrightarrow CH_2O + h\nu$$

Emission is centred on 435 nm and hence no interference from the reaction of ozone and nitrogen oxide occurs. The method is sensitive to as little as 1 ppb of ozone.

The uv absorption of ozone at 254 nm may be used for its determination at levels down to 3 ppb. Interferences from other uv-absorbing air pollutants such as mercury and hydrocarbons may be eliminated by taking two readings. The first reading is of the absorbance of an air sample after catalytic conversion of ozone to oxygen and the second is of an unchanged air sample, the difference in absorbance being due to the ozone content of the air. Available instruments perform this procedure automatically and give read-out in digital form. Although truly continuous measurement of ozone levels is not possible, response is fast and readings may be taken at intervals of less than one minute.

2.5.2 Peroxyacetyl Nitrate (PAN)

PAN is a product of atmospheric photochemical reactions and is characteristic of photochemical smog.

$$CH_3 - \underset{\underset{O}{\|}}{C} - O - O - NO_2 \qquad \text{PAN}$$

Levels in Southern California lie typically within the range 5-50 ppb on smog days. In Europe, the formation is far less favoured and concentrations are more usually <10 ppb.

2.5.2.1 Chemical Analysis

PAN may be determined by long-path ir measurements, the greatest sensitivity being achieved when Fourier transform methods are used. The most sensitive and specific routine technique involves gas chromatographic separation and detection of specific peroxyacyl nitrates by electron capture. The detection limit of below 1 ppb permits it to be used as a direct atmospheric monitor under circumstances of high pollution.

2.6 Particulate Pollutants

These contain a great diversity of both water-soluble and insoluble components. About 30-40% typically is insoluble in water and comprises man-made materials such as elemental and organic carbon and corrosion-derived iron oxides, as well as natural mineral materials such as quartz, calcite and clay minerals derived from surface erosion processes.

The water-soluble material consists of nine major ionic components: Na^+, K^+, NH_4^+, Ca^{2+}, Mg^{2+}, H^+, SO_4^{2-}, NO_3^-, Cl^- which in total comprise about 60-70% of the total aerosol mass. The major proportion of the Na^+, Mg^{2+} and Cl^- arises from marine sources, whilst Ca^{2+} and K^+ are mainly soil derived. The major source of the other components, NH_4^+, H^+, SO_4^{2-} and NO_3^-, is the atmospheric oxidation of SO_2 and NO_x to sulphuric and nitric acids, and their total or partial neutralisation by natural atmospheric ammonia.

The main source of primary man-made particulate pollutants is the combustion of fossil fuels. Table 7 shows the percentages arising from each type of fuel used.

Table 7 U.K. Primary particulate emissions from fuel combustion

	1977		1980	
	Thousand tonnes	Per centage	Thousand tonnes	Per centage
Coal	398	72	306	69
Solid smokeless fuel	44	8	35	8
Petroleum				
Motor spirit	26	5	29	6
Derv	34	6	35	8
Gas oil	19	3	16	4
Fuel oil	23	4	16	4
Refinery fuel	5	1	5	1
All fuels	549	100	442	100

It will be appreciated that much additional particulate aerosol material is natural (e.g. the soil-derived and marine components) and much is secondary, arising as gas to particle conversion products (most notably H_2SO_4, NH_4HSO_4, $(NH_4)_2SO_4$ and NH_4NO_3). Other, minor but important components of atmospheric aerosol are the toxic metal pollutants. Respiratory exposure can be an important route of intake for some metals such as lead and cadmium.

The particle loading of the air may be measured by a number of techniques. The two most important methods measure different quantities and give rise to separate definitions. Smoke is defined as fine suspended particulate air pollutants (< 15 μm) as measured by determining the staining capacity of the air, a major contribution over the country as a whole being the incomplete combustion of coal. Total suspended particulates are particles (< 15 μm) suspended in the atmosphere, as collected by filtration with subsequent gravimetric determination, irrespective of the relative staining capacities of these emissions.

Analysis of specific ionic components of atmospheric particles may be carried out by leaching into a suitable solvent, followed by a chromatographic or spectrophotometric determination. Metallic substances are normally dissolved in strong oxidising acids and analysed by atomic absorption spectrophotometry or anodic stripping voltammetry. Direct determination of metals on the surface of an air filter is also possible by X-ray fluorescence or instrumental neutron activation analysis.

An important organic component of combustion-derived particulates is the polynuclear aromatic hydrocarbons (PAH), some of which are carcinogenic in experimental animals.

After solvent extraction, PAH are determined by column chromatographic separation and uv absorbance measurement. Recently, application of gas-liquid chromatography (glc) with flame ionisation or electron capture detection, and of high pressure liquid chromatography with detection by uv absorption or fluorescence measurement, has reduced detection limits from the microgram to the nanogram level or below, for individual compounds. Whilst the glc separation of highly carcinogenic benzo(a)pyrene from its less carcinogenic isomer benzo(e)pyrene has given problems, reasonable resolution may be achieved. A more positive identification of glc peaks has been given by gc-ms.

Suspended particles may be a human health hazard due to their chemical content (e.g. toxic metals, carcinogenic PAH), or due to a synergistic effect with sulphur dioxide. Elevated levels of suspended matter are associated with reduced visibility in the atmosphere, this apparently being linked to the sulphate content of the air. The U.S.E.P.A. ambient air quality standard for total suspended particulates is 75 $\mu g\ m^{-3}$ (annual geometric mean) and 260 $\mu g\ m^{-3}$ (24h average, not to be exceeded more than once per year).

3. BIBLIOGRAPHY

General text books

H.C. Perkins, "Air Pollution", McGraw-Hill, New York, 1974.

A.C. Stern, H.C. Wohlers, R.W. Boubel and W.P. Lowry, "Fundamentals of Air Pollution", Academic Press, New York, 1973.

Emission and air quality data

Data, derived mainly from the Warren Spring Laboratory, and used in this article are given in:

Department of the Environment, "Digest of Environmental Pollution and Water Statistics", No 4, H.M.S.O., London, 1981.

More specific articles

R.M. Harrison and C.D. Holman, Ozone Pollution in Britain, Chemistry in Britain, 18, 563-70 (1982).

R. Perry and R.M. Harrison, Air, A Measure of Pollution, Chemistry in Britain, 12, 185-98 (1976).

R.M. Harrison and H.A. McCartney, An Automated Mobile Laboratory for Continuous Monitoring of Atmospheric Pollutants, Trans. Inst. Mech. Eng., 194, 357-64 (1980).

4. APPENDIX

Air Pollutant Concentration Units

Probably the most logical unit of air pollutant concentration is mass per unit mass, i.e. $\mu g\ kg^{-1}$ or $mg\ kg^{-1}$. This is, however, very rarely used. The commonest units are mass per unit volume (usually $\mu g\ m^{-3}$) or volume per unit volume, otherwise known as a volume mixing ratio (ppm or ppb). For particulate pollutants the volume mixing ratio is inapplicable.

Much confusion arises in the interconversion of $\mu g\ m^{-3}$ and ppm. Whilst the volume mixing ratio is independent of temperature and pressure for an ideal gas (and air pollutant behaviour is close to ideal), the mass per unit volume unit is dependent on T and P conditions, and hence these will be taken into account.

Example 1. Convert 0.1 ppm nitrogen dioxide to $\mu g\ m^{-3}$ at $20^{o}C$ and 750 torr.

46 g NO_2 occupy 22.41 ℓ at STP

46 g NO_2 occupy $22.41 \times \frac{293}{273} \times \frac{760}{750}\ \ell$

$= 24.37\ \ell$ at $20^{o}C$ and 750 torr

0.1 ppm NO_2 is $10^{-7}\ \ell\ NO_2$ in 1 litre, or

. . . . $10^{-4}\ \ell\ NO_2$ in 1 m^3

$10^{-4}\ \ell\ NO_2$ at $20^{o}C$ and 750 torr contain $46 \times \frac{10^{-4}}{24.37}$ g

$= 189\ \mu g\ NO_2$

$\therefore$ NO_2 concentration = $\underline{189\ \mu g\ m^{-3}}$

Example 2. Convert 100 $\mu g\ m^{-3}$ ozone at $25^{o}C$ and 765 torr to ppb.

48 g ozone occupy 22.41 ℓ at STP

. . . . . occupy $22.41 \times \frac{298}{273} \times \frac{760}{765}\ \ell$

$= 24.30\ \ell$ at $25^{o}C$ and 765 torr

100 μg ozone occupy $24.30 \times \frac{100 \times 10^{-6}}{48}\ \ell$

$= 50.6 \times 10^{-6}\ \ell$ at $25^{o}C$ and 765 torr

$\therefore$ Volume mixing ratio $= 50.6 \times 10^{-6} (\ell) \div 1000 (\ell)$

$= 50.6 \times 10^{-9}$

$= \underline{51\ ppb}$

9
Pollutant Pathways in the Atmosphere

By R. J. Donovan
DEPARTMENT OF CHEMISTRY, UNIVERSITY OF EDINBURGH, WEST MAINS ROAD, EDINBURGH EH9 3JJ, U.K.

1. Introduction

In this chapter we shall examine the detailed chemical pathways that lead to the removal of gaseous pollutants from the atmosphere. A clear account of the sources of gaseous pollutants, their magnitude and the analytical techniques used to monitor pollutants has been given in the previous chapter and this will be taken as our starting point. It is important to recognise at the outset that the chemistry of the atmosphere is controlled largely by atoms and free-radical species that are present in very low concentrations, typically in the region of 10^7 cm^{-3} (equivalent to 10^{-12} atm.). It is the highly reactive nature of free-radical species that leads both to their low concentration and their importance as intermediates in atmospheric chemistry. Two further consequences are that free radicals are difficult to study in the laboratory and even more difficult to detect at ambient concentrations in the atmosphere. These two difficulties are the main source of the uncertainties which exist in some of the pathways to be discussed below.

Before embarking on a detailed discussion of chemical removal mechanisms it is essential to have at least a general appreciation of the structure of the atmosphere. The temperature profile and variation in density with altitude are shown in figure 1, together with the nomenclature for the various regions (based on the temperature profile). We shall be mainly concerned here with the troposphere,

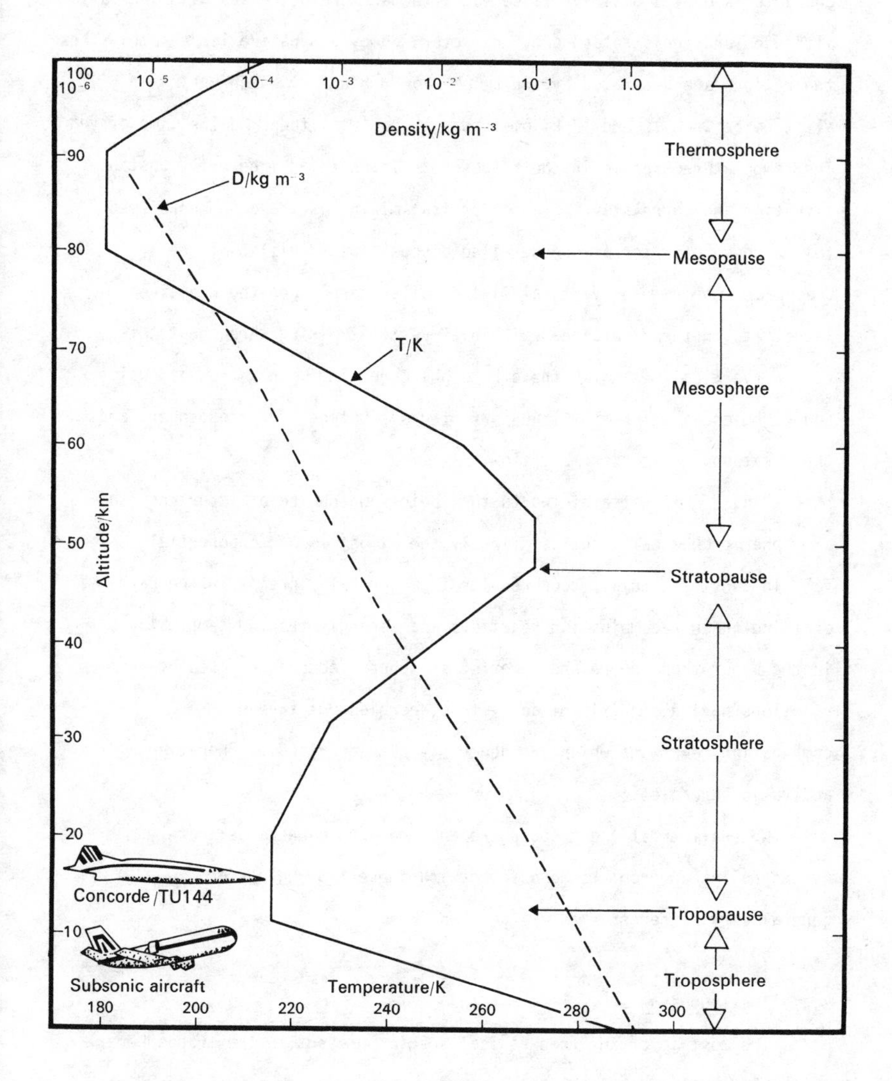

Fig. 1. Physical structure and nomenclature of the atmosphere.

characterised by a positive lapse rate (temperature decreases with increasing altitude) and the stratosphere, characterised by a negative lapse rate. These two regions are separated by the tropopause, a boundary which varies in altitude between 10 and 17 kilometres with both latitude and the time of year. The temperature profile in the troposphere leads to an inherently unstable situation and to relatively rapid vertical mixing of the components (see following chapter for a more detailed discussion of this). At the tropopause the rate of vertical mixing falls sharply and the negative lapse rate in the stratosphere inhibits vertical mixing throughout this region (vertical mixing in the stratosphere results mainly from diffusion and turbulence). Thus the transport of gases between the troposphere and the stratosphere is extremely slow.

Figure 1 illustrates two further points which are of importance to atmospheric chemical kinetics: firstly the almost smooth exponential fall in density with altitude results in a general decrease in the rate of third-order recombination reactions and secondly the low temperatures in the upper troposphere and lower stratosphere reduce the rates of reactions having significant activation energies but favour combination reactions which are generally characterised by "negative" activation energies.

We shall treat the troposphere and the stratosphere separately as the pollution problems associated with these two regions are in general very different.

2. The Troposphere

The most important free-radical species present in the troposphere is the hydroxyl radical which is formed by the reaction of O^1D (the first electronically excited state of the oxygen atom) with water vapour, viz.

$$O(^1D) + H_2O \rightarrow 2OH \quad (1)$$

the oxygen ^1D atom is formed by photolysis of ozone:

$$O_3 + h\nu\ (\lambda<320\ nm) \rightarrow O(^1D) + O_2(^1\Delta) \quad (2)$$

Most of the O^1D is rapidly quenched by O_2 or N_2 in the atmosphere and this greatly reduces the chance of reaction with water; however, quenching simply returns oxygen atoms to O_3 by the reaction

$$O(^3P) + O_2 + M \rightarrow O_3 + M \quad (3)$$

The O_3 is then photolysed again giving O^1D and the cycle is repeated until O^1D meets a water molecule and forms OH.

We shall now consider the reactions of the OH radical in terms of the pollutants discussed in the previous chapter.

Hydrocarbons

Hydrocarbons are removed by what is essentially a low temperature combustion process the end products being CO_2 and water. This is illustrated by considering the removal of CH_4 (see figure 2). The initial step involves attack by a hydroxyl radical, or to a lesser extent O^1D, to form a methyl radical. This step is slow for methane due to the relatively high CH bond strength but becomes fairly rapid for hydrocarbons with secondary or tertiary hydrogen atoms. The resulting methyl radical then reacts almost instantaneously with O_2 to form a methyl peroxy radical which is relatively stable. The next step is again relatively slow and involves reaction with NO, or photolysis, to yield a methoxy radical. Attack by O_2 then removes a further hydrogen atom to yield formaldehyde. The formaldehyde is either photolysed or reacts with hydroxyl followed by O_2 to yield carbon monoxide. The final step is reaction of carbon monoxide with hydroxyl to yield CO_2. This last step is a particularly important one and is the prime mechanism for removing carbon monoxide, from all sources, from our atmosphere. The reaction is rapid and has little or no activation energy. The resulting hydrogen atom goes on to combine with an O_2 molecule and forms the HO_2 radical which itself has an interesting chemistry, particularly in the stratosphere.

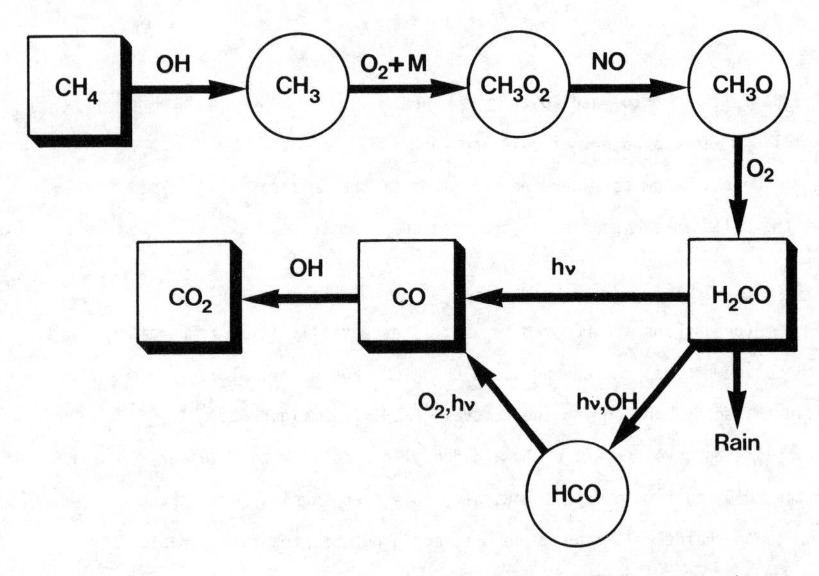

Figure 2. Reaction paths leading to the removal of CH_4 from the atmosphere.

Most of the steps in this mechanism have been studied in considerable detail and it can now be considered as relatively well established. Some uncertainty remains over the precise rate of some of the steps and further work to refine these would be valuable.

Sulphur Dioxide

Sulphur dioxide is a fairly soluble gas and under wet conditions can be removed as sulphurous acid or by reactions in the liquid phase (i.e. rain drops, etc.). However, dry removal must also be important and here the rate-determining step is thought to be oxidation to form SO_3 (sulphur trioxide is known to react rapidly with water to form sulphuric acid). Three mechanisms have been proposed for this oxidation step, the first involving the combination of SO_2 with an oxygen atom;

$$SO_2 + O + M \rightarrow SO_3 + M \qquad (4)$$

The rate constant for this reaction is known; however, our knowledge of the free oxygen atom concentration in the troposphere is less well established. With an oxygen atom concentration of ca. 10^5 cm^{-3} the half life of SO_2 would be approximately 10^2 days. This is two slow to account for the main bulk of SO_2 removal, particularly as the oxygen atom concentration is probably lower than the figure taken, but this mechanism must clearly play a small part in the total removal.

A second mechanism involving photo-oxidation of SO_2 has also been proposed. This involves the excitation of SO_2 by sunlight to a singlet state from which it is collisionally quenched into a nearby metastable triplet state. It has been proposed that the energised and metastable SO_2 then combines with O_2 to form SO_4 and the SO_4 species reacts with a further O_2 yielding SO_3 and ozone. Unfortunately SO_4 has not been positively identified in laboratory experiments and the rates of the final two steps are not known. It is difficult therefore to assess quantitatively the importance of this scheme in the removal of SO_2.

The most plausible scheme for the dry removal of SO_2 again involves the OH radical in the reaction given below,

$$OH + SO_2 + M \rightarrow HSO_3 + M \qquad (5)$$

The rate of this reaction is now established and it appears that it can fully account for the removal of SO_2 provided removal of HSO_3 is rapid.

NO_x Species

As mentioned in the previous chapter nitrous oxide (N_2O) is relatively inert in the troposphere. This has consequences in terms of stratospheric pollution which we shall consider in the next main section. Nitric oxide (NO) and nitrogen dioxide (NO_2) must both be considered together as they are rapidly interconverted in the troposphere. During daylight conditions NO_2 is photolysed to yield an oxygen atom and NO. The oxygen atom then combines with O_2 to form ozone which can return to react with NO and form NO_2 again. Thus under bright sunlight conditions the main species present in the atmosphere will be NO while under low light level conditions the main species will be NO_2.

The main removal process for NO_x species in the troposphere is again a reaction involving OH radicals, viz:

$$OH + NO_2 + M \rightarrow HNO_3 + M$$

The nitric acid is then removed by rain.

The reactions of NO_x in the troposphere are well illustrated by their involvement in photochemical smog formation and we shall therefore discuss this next. The now classical photochemical smog cycle and the variation in the chemical species involved with time are illustrated in figure 3.

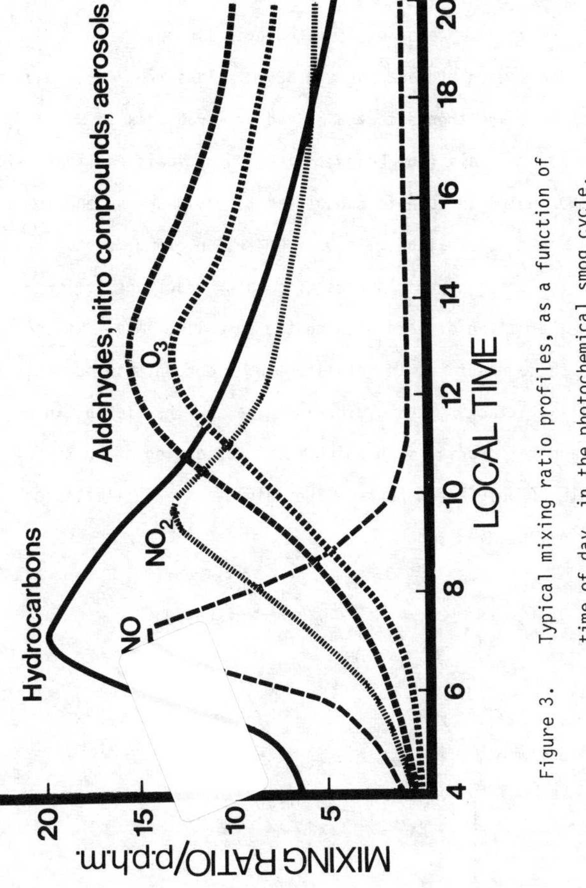

Figure 3. Typical mixing ratio profiles, as a function of time of day, in the photochemical smog cycle.

The four main requirements for photochemical smog formation are:

1. strong sunlight
2. stable meteorological conditions
3. the presence of NO_x
4. the presence of unsaturated hydrocarbons.

The cycle starts with hydrocarbon and NO emissions from car exhausts. The hydrocarbons are then attacked by hydroxyl radicals to yield eventually RO_2 radicals (see below). The RO_2 radicals can then oxidise the NO to NO_2 which in turn is photolysed to yield an oxygen atom and ultimately, on combination with O_2, forms an ozone molecule. Once ozone is formed a rapid series of complex reactions follow which lead to the formation of aerosols and the so-called smog. We shall not follow these reactions in detail but will concentrate our attention on the initial attack by the hydroxyl radical on the olefin, in order to understand the crucial step which is the oxidation of NO to NO_2 on a time scale of only a few hours. The main steps are illustrated in the following reaction scheme:

$$R_2C{=}CH_2 \xrightarrow{OH} R_2C(OH){-}\dot{C}H_2 \xrightarrow{O_2} R_2C(OH){-}CH_2O_2\cdot$$

$$R_2C(OH){-}CH_2O_2\cdot \xrightarrow{NO} R_2C(OH){-}CH_2O\cdot + NO_2$$

$$R_2C(OH){-}CH_2O\cdot \longrightarrow CH_2O + R_2\dot{C}(OH)$$

$$R_2\dot{C}(OH) \xrightarrow{O_2} R_2C(OH)O_2\cdot$$

$$R_2C(OH)O_2\cdot \xrightarrow{NO} R_2C(OH)O\cdot + NO_2$$

3. The Stratosphere

The stratosphere is a particularly vulnerable region of the atmosphere due to its low density and its stability against vertical mixing. First concern over the pollution of the stratosphere arose in about 1970 when it was realised that flights of supersonic aircraft could inject appreciable quantities of exhaust gases directly into the lower stratosphere and thereby possibly reduce the ozone concentration. The main effect of this would be to increase the levels of ultraviolet light reaching ground level (mainly in the uv-B region, 290-320 nm) and would cause undesirable biological effects including a possible increase in skin cancer. Attention was at first focussed on the effects of water vapour but it was soon realised that the oxides of nitrogen, present in the exhaust gases, would play a much more important role. It should be emphasised that at the time of this debate it was thought that there would be several fleets of supersonic aircraft operating in the 1980's. This has clearly not happened owing to the early cancellation of the Boeing SST project and the commercial failure of the Soviet TU144 and Anglo-French Concordes. The effect of such flights on the stratosphere at present is very much less than that of conventional aircraft; however the effort that went into the assessment of this area was not wasted as it has lead to a quantitative understanding of NO_x pollution of the stratosphere.

NO_x Species

Before discussing pollution by NO_x species we must first look at the natural background NO_x that is present in the stratosphere. The lack of sinks for N_2O in the troposphere means that its main sink is by slow transport into the stratosphere where it is photolysed or undergoes reaction with O^1D to form nitric oxide. The photolysis of N_2O which is the main sink in the stratosphere produces molecular nitrogen and an oxygen atom which is of little significance in the overall chemistry of the stratosphere. However, a minor process involving attack by O^1D is of considerable significance as the product, nitric oxide, enters directly into what is known as the NO_x cycle. This highly efficient catalytic cycle has a strong influence on the ambient ozone concentration in the lower stratosphere. The two important steps in the NO_x cycle are illustrated at the top of figure 4. In the first step NO reacts with ozone to form NO_2. The NO_2 is then attacked by an oxygen atom, releasing nitric oxide which returns to the first step in the cycle. The combined effect of these two steps is to consume oxygen atoms and O_3 converting them into molecular oxygen. Both oxygen atoms and O_3 are termed <u>odd oxygen</u> and they can be viewed as chemically equivalent in the stratosphere as oxygen atoms ultimately combine with O_2 to form O_3 while the photolysis of O_3 releases oxygen atoms. Thus, throughout the stratosphere O_3 and oxygen atoms are rapidly interconverted and can essentially be regarded as equivalent. The NO_x cycle accounts for about 60% of the ozone removal rate in the lower startosphere and is clearly an important process. The important point to grasp is that the reaction between oxygen atoms and O_3 is unusually slow as it has an activation energy of 19 kJ per mole. Thus the role of NO_x is to catalyse the reaction and to greatly increase the overall rate.

The main sink in the natural NO_x cycle is also illustrated in figure 4. The hydroxyl radical is again involved in the formation of nitric acid which is slowly transported into the troposphere where it is rained out.

It is clear from the above discussion that the introduction of additional quantities of nitrogen oxides into the stratosphere will lead to a reduction in the ozone concentration. The concern over supersonic aircraft was that they would inject NO and NO_2 directly into the lower

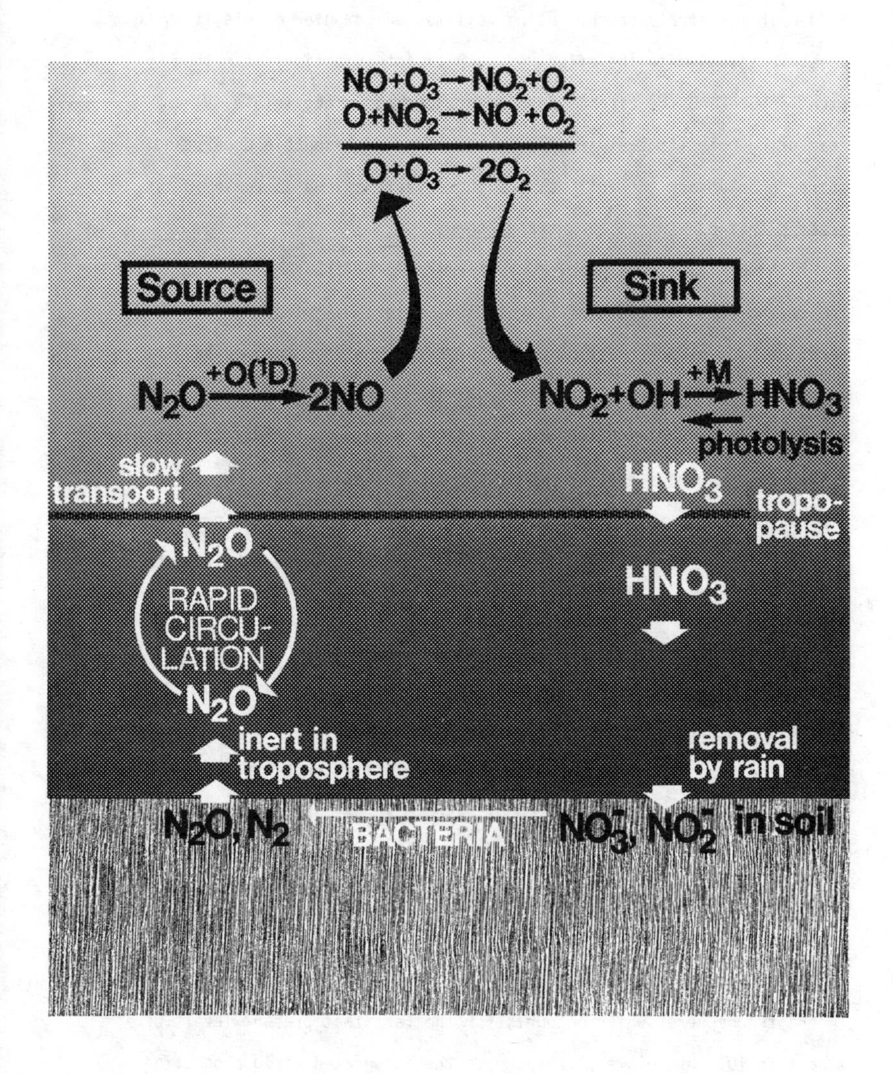

Figure 4. Main features of the NO_x cycle with sources and sinks.

stratosphere and thus enhance the natural NO_x cycle. Subsonic aircraft also penetrate the lower stratosphere on Polar flights and add to the NO_x burden in the stratosphere. Using well tested computer models it is now possible to estimate the effect of such direct injections into the stratosphere and the effects have been shown to be negligible. The reassuring point is that in this case we have a reasonable understanding of the problem and can fully assess the effects of any increase in air traffic that penetrates the stratosphere.

A more difficult problem has been posed by the suggestion that the increasing world wide use of nitrate fertilisers will lead to increased rates of N_2O release. Estimates of the effect on the ozone cover vary widely reflecting our poor knowledge of the natural N_2O cycle and this problem cannot in fact be resolved until we have a more detailed knowledge of the sources and sinks for N_2O in the troposphere. In the meantime we are faced with an interesting moral problem: should we try to prevent any increase in the use of nitrate fertilisers - a move unlikely to be popular with developing countries - or run the risk of producing significant changes in the ozone cover?

Chlorofluorocarbons

The possible effects of stratospheric pollution by the chlorofluorocarbons (CFCS) have been widely publicised. These materials were used in increasing quantities throughout the 60's and early 70's as aerosol spray propellants, refrigerants, solvents and plastic foaming agents and have been accumulating steadily in the atmosphere (mainly $CFCl_3$ and CF_2Cl_2). It had generally been assumed that CFC's were inert, as indeed they are in the troposphere, which accounts for the fact that present levels of 1 part in 10^{10} by volume correspond to the integrated world production to date. The problem arises once they are transported into the stratosphere where they are photolysed or react with electronically excited oxygen atoms producing chlorine atoms and chlorine oxides (collectively termed ClO_x). These degradation products form the basis of a further catalytic cycle, analogous to the NO_x cycle, which leads to the net removal of ozone and oxygen atoms, viz.

$$Cl + O_3 \rightarrow ClO + O_2 \quad (6)$$

$$ClO + O \rightarrow Cl + O_2 \quad (7)$$

$$O + O_3 \rightarrow 2O_2 \quad (8)$$

Again, the overall rate for the catalysed reaction is much greater than that between O and O_3. Model calculations based on the available rate data for these reactions have suggested that the continuing use of CFC's at current levels could lead to a reduction of between 5 and 9% in the ozone coverage after about 100 years (a period of approximately 100 years is required to reach steady state as a consequence of the very slow rate of transport into the stratosphere). However, this figure for the overall ozone depletion remains very uncertain and has changed quite dramatically over the last few years as new rate data and chemical reaction pathways have been incorporated into the models. One example of a reaction that caused a major revision is the combination of ClO with NO_2 to form chlorine nitrate, viz.

$$ClO + NO_2 \xrightarrow{+M} ClNO_3 \quad (9)$$

This provides a sink for both ClO_x and NO_x and thereby reduces the efficiency of both cycles. The late discovery of reactions of this type raises the question of whether there could also be other Cl-O-H-N containing compounds present in the atmosphere which have not been included in the model calculations and could thus change the estimates for net ozone depletion.

Another uncertainty is the quantitative assessment of natural ClO_x sources. It appears that CH_3Cl has considerable natural sources and that its flux into the stratosphere, together with naturally released CCl_4, is comparable to the present flux of CFC's. Current efforts to obtain *in situ* measurements on a wide range of chlorine-containing species in the stratosphere should help to resolve the uncertainty concerning the pollution of the stratosphere by CFC's.

An important difference between pollution by CFC's and N_2O on the one hand and aircraft on the other is the time taken for the stratosphere to recover. For the NO_2 and NO, injected directly into the stratosphere by aircraft, recovery occurs within approximately 3 years. However, the release of CFC's and N_2O involves much longer recovery times due to the buffer effect of the large mass of the troposphere (approximately 10 times that of the stratosphere) and the slow rate of transport from the troposphere into the stratosphere. The half recovery time with CFC's, assuming that release was completely terminated, has been estimated to be between 30 and 50 years.

Further Reading

I.M. Campbell, "Energy and the Atmosphere", Wiley and Sons, 1977.

R.A. Cox and R.G. Derwent, "Gas-phase Chemistry of the Minor Constituents of the Troposphere", in Specialist Periodical Reports of the Royal Society of Chemistry, Gas Kinetics and Energy Transfer, Vol. 4 (1981).

H.S. Johnstone, Ann. Rev. Phys. Chem. 26, 315 (1975).

B.A. Thrush, Endeavour 1, 3 (1977).

10

Atomspheric Dispersal of Pollutants and the Modelling of Air Pollution

By A. W. C. Keddie
AIR POLLUTION DIVISION, WARREN SPRING LABORATORY, GUNNELS WOOD ROAD, STEVENAGE, HERTS. SG1 2BX, U.K.

1. Introduction

Considerable resources are devoted to the measurement of air pollutant concentrations in the ambient atmosphere but measurements on their own provide little information on the origin of the pollutants in question, on the dispersal process in the atmosphere and on the impact of new sources or the benefits of controls. There is frequently the need, therefore, for detailed knowledge of the characteristics and quantities of pollutants emitted to the atmosphere and on the atmospheric processes which govern their subsequent dispersal and fate. This knowledge must then be built into an appropriate dispersion model, whether the problem to be addressed is the emissions from a single chimney or the emissions from a large multi-source urban/industrial area. The 1956 Clean Air Act Chimney Height Memorandum[1] is an example of what is essentially a model for single chimneys whilst the work carried out by WSL in the Forth Valley of Scotland[2] provides an example of a multi-source model.

The trend towards formal air quality guidelines or standards (e.g. as recently agreed within the EEC for smoke and sulphur dioxide[3]) is creating the need for formal air quality management systems and for a more strategic approach to air pollution control. Neither can be accomplished without the use of atmospheric dispersion modelling techniques. A wide variety of techniques are available, ranging from the most simple "box" model through to numerical solutions of the basic equations of fluid flow, etc. For the more chemically reactive pollutants, it is also necessary to incorporate the relevant atmospheric chemistry. The spatial and temporal resolution and accuracy of the model output ideally must match the questions being posed. Where emissions vary greatly, both in space and time, or it is necessary to predict the time series of concentrations at specified locations, then the modelling task is extremely difficult, even without the complications of a very chemically reactive pollutant species or substantial topographical effects on the dispersal pattern. On the other hand, if it is not necessary to know when a specified concentration will occur but rather it is the probability of occurrence during a given period (e.g. a year) which is of interest, then the modelling task is generally much less demanding. For most regulatory purposes and for the strategic planning of air quality it is the probability of occurrence which matters; for example, an air quality standard may be defined in terms of limits on the median and 98th percentile of the annual cumulative frequency distribution[3]. This lecture is primarily concerned with this task for relatively non-reactive pollutants, such as sulphur dioxide and suspended particulates, over distance scales of a few hundred metres to about 50km.

2. Atmospheric Dispersion

Material discharged into the atmosphere is carried along by the wind and diluted by the turbulence which is always present. This dispersal process has the effect of producing a plume of polluted air which, to a first approximation, is roughly cone shaped with the apex towards the stack. Fig. 1 illustrates this much simplified concept of a plume and mathematically it can be described[4,5] by the Gaussian equation:

$$\chi_{(x,y,z)} = \frac{Q}{2\pi \sigma_y \sigma_z U} \exp\left[-\frac{1}{2}\left(\frac{y}{\sigma_y}\right)^2\right]\left\{\exp\left[-\frac{1}{2}\left(\frac{z-H_e}{\sigma_z}\right)^2\right] + \exp\left[-\frac{1}{2}\left(\frac{z+H_e}{\sigma_z}\right)^2\right]\right\} \quad (1)$$

where χ is the pollutant concentration at point (x, y, z), $\mu g\ m^{-3}$
Q is the pollutant emission rate, $\mu g\ s^{-1}$
U is the wind speed, $m\ s^{-1}$
σ_z is the standard deviation of the plume concentration in the vertical at distance x, m
σ_y is the standard deviation of the plume concentration in the horizontal at distance x, m
x, y and z are the along wind, crosswind and vertical distances (m) with the base of the stack as the origin for the distance co-ordinates
H_e is the effective height (stack height plus buoyancy or momentum rise) of the plume, m

With y = z = 0 equation (1) reduces to the familiar equation for the ground level concentration below the plume centre line:

$$\chi_{(x)} = \frac{Q}{\pi \sigma_y \sigma_z U} \exp\left[-\frac{1}{2}\left(\frac{H_e}{\sigma_z}\right)^2\right] \quad (2)$$

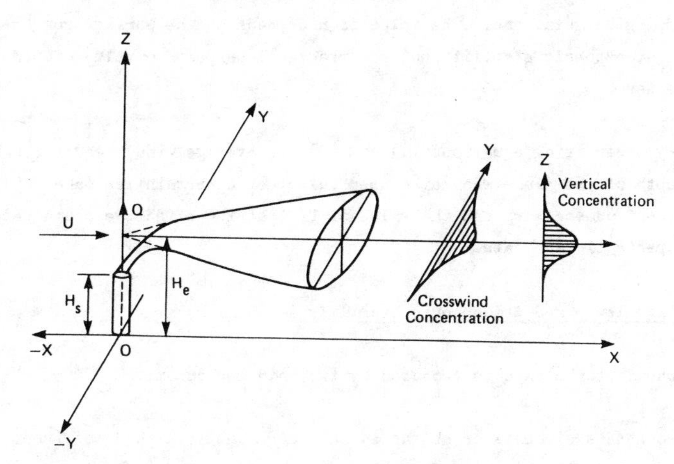

Figure 1. *Gaussian Plume Distribution*

Rate of growth of a plume, its dimensions and its shape, and therefore the parameters in equations (1) and (2) are determined by the wind speed, the degree of turbulence and in many cases the presence of buildings or topographical features.

2.1 Wind Speed

In the lower layers of the atmosphere wind speed increases with increasing height above ground because the effect of the drag or frictional forces at the surface of the earth diminishes with height. For many practical purposes it is convenient to describe the wind speed profile by the power law:

$$U_{(z)}=U_{10}(z/10)^{n} \tag{3}$$

where $U_{(z)}$ is the wind speed at height z and U_{10} is the wind speed at the reference height of 10 metres; U_{10} is normally what is available from meteorological stations. The value of n depends on the surface roughness and on atmospheric stability but for practical purposes a value of 0.2 can be used.

Ideally, when using equations (1) and (2) the average wind speed through the depth of the plume should be used but other uncertainties make this refinement unnecessary and the value of $U_{(z)}$ at the effective plume height, H_e, is perfectly adequate.

2.2 Turbulence and Atmospheric Stability

Atmospheric turbulence is produced by two main mechanisms;

(i) Roughness elements or obstacles on the ground (e.g. hedges, trees, buildings and hills) engender mechanical mixing of the air as it passes over them. The degree or intensity of mechanical turbulence increases with increasing surface roughness and increasing wind speed and diminishes with increasing height above the ground.

(ii) Solar radiation heats the surface of the earth, thus increasing the temperature in the lower atmosphere and creating density differencies, which generate upward buoyancy forces. The motion of individual parcels of air is unstable, these turbulent motions being most intense during hot sunny weather with light winds. At night when there is no incoming solar radiation and the surface of the earth cools, temperature increases with height and turbulence tends to be suppressed. During calm clear nights when surface cooling is rapid and little or no mechanical turbulence is being generated, turbulence may be almost entirely absent.

When considering most dispersion problems it is convenient to classify the possible states of the atmosphere into what are usually referred to as stability categories. The typing sceme developed by Smith from the original Pasquill formulation[4,5] is widely used because of its relative simplicity yet dependence on sound physical principles. Stability is classified according to the amount of incoming solar radiation, wind speed and cloud cover. A semi-quantitative guide is given in Table 1 and

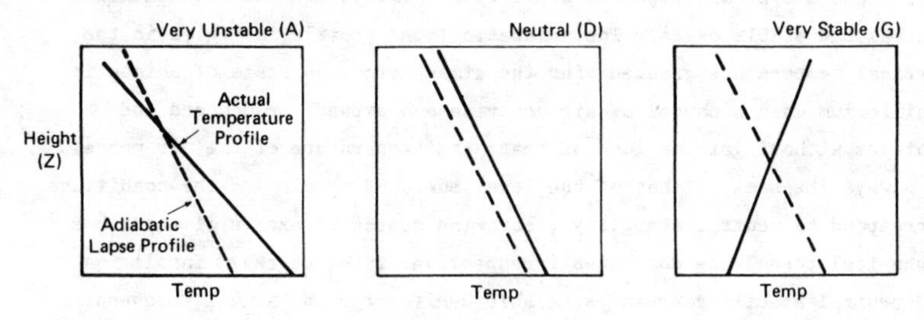

Figure 2. Typical Atmospheric Temperature Profiles and Corresponding Stabilities

TABLE 1 – Pasquill's Stability Categories*

Surface wind speed m s⁻¹ (≡U₁₀)	Insolation			Night	
	Strong	Moderate	Slight	Thickly overcast or ≥4/8 low cloud	≤3/8 cloud
<2	A	A–B	B	–	G
2–3	A–B	B	C	E	F
3–5	B	B–C	C	D	E
5–6	C	C–D	D	D	D
>6	C	D	D	D	D

Strong insolation corresponds to sunny midday in midsummer in England, slight insolation to similar conditions in midwinter. Night refers to the period from 1 hour before sunset to 1 hour after dawn. A is the most unstable category and G the most stable. D is referred to as the neutral category and should be used, regardless of wind speed, for overcast conditions during day or night.

* Based on Fig. 6.10 in reference 4.

TABLE 2 – Typical Annual Frequency of Occurrence of Stability Categories in Great Britain

Stability Category	Frequency of Occurrence %
A	0.6
B	6.0
C	17.0
D	60.0
E	7.0
F	8.0
G	1.4

Fig. 2 shows typical temperature profiles corresponding to the unstable, neutral and stable cases. The adiabatic lapse profile in Fig. 2 is the vertical temperature gradient for the atmosphere in a state of adiabatic equilibrium when a parcel of air can rise and expand, or descend and contract without gain or loss of heat; the temperature of the air parcel is always the same as that of the level surrounding air and the conditions correspond to neutral stability. For wind speeds in excess of about $6 ms^{-1}$ mechanical turbulence dominates irrespective of the degree of insolation and neutral stability prevails. Table 2 gives typical annual frequencies of occurrence of the different stability categories in Great Britain.

Turbulent motions or eddies in the atmosphere vary in size and intensity; the greater their size and/or intensity the more rapid is the plume growth and hence the dilution of the pollutants. Small scale turbulent motions tend to dominate the plume growth close to the point of emission where the plume is still relatively small, and the larger scale eddies dominate at greater distances. Furthermore, the small and larger eddies are associated respectively with short and larger time scales. Consequently, σ_y and σ_z increase in value with distance from the source (see Fig 1); also they increase with the time or sampling period over which they have been measured. This latter point means that it is essential to state the sampling period to which σ_y and σ_z apply, especially if comparisons are being made between calculated and measured concentration; ideally the two periods should be identical. Most values of σ_y and σ_z to be found in the literature are for sampling periods in the range 3-60 minutes. It is also evident that σ_y and σ_z are dependent on atmospheric stability, being smallest when the atmosphere is most stable (category G), i.e. when atmospheric turbulence is least, increasing to their greatest values in highly turbulent very unstable conditions (category A). The underlying surface roughness elements also play a part, σ_y and σ_z increasing with increasing surface roughness so that for a given distance downwind of a chimney σ_y and σ_z will be larger in, for example, an urban area than in an area of open relatively flat agricultural land.

In general, lateral (horizontal) motion is less constrained than verticle motion with the result that there are larger scale eddies in the horizontal than in the vertical. Fluctuations in wind direction also become important for longer sampling periods. Consequently, σ_y increases more rapidly with increasing sampling or averaging period than does σ_z. This dependence of σ_y on wind direction fluctuations also means that for longer sampling periods, say greater than 1 hour, σ_y values can increase with increasing atmospheric stability because during low wind speed stable conditions plume meandering can be significant.

2.3 Mixing Heights

Atmospheric stability often exhibits considerable variation with height, e.g. an elevated stable inversion layer (temperature increasing with height) capping a surface unstable layer, or a neutral regime above a ground based temperature inversion. The height of the boundary between two such stability regimes is usually referred to as the mixing height. The vertical dispersion (dilution) of pollutants emitted below an elevated inversion will be inhibited whilst emissions above or within the inversion may be effectively prevented from reaching the ground, both circumstances having a significant influence on resultant ground level concentrations. Quantitative information on the heights of such barriers, their frequency of occurrence and their variability throughout the day (e.g. a ground-based inversion will be generally eliminated by solar heating during the morning) is therefore essential. At WSL we have made extensive use of SODAR (sound detection and ranging) to obtain this information[6,7]. Figure 3 is an example of how mixing heights vary with atmospheric stability, derived from SODAR data in Central Scotland[7].

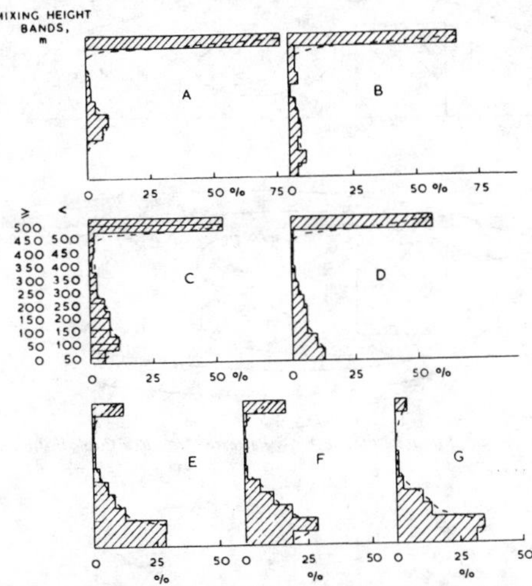

Fig. 3 Frequency distributions of mixing height bands by stability category (May 1976–April 1977).

2.4 Building and Topographical Effects

Hills or buildings can have significant adverse effects on plume dispersion if their dimensions are large in comparison with the dimensions of the plume or if they significantly deflect or disturb the flow of the wind. Figure 4 shows a simplified and idealised representation of the flow over a building. There is a zone on the immediate downwind (or leeward) side of the building which is to some extent isolated from the main flow and within which there is a reversal of the air flow. Further downstream the air flow is highly turbulent. Waste gases escaping through a relatively short chimney,attached or adjacent to the building, will be entrained in this characteristic flow pattern and will not disperse according to the conventional equations (1) and (2). Recent wind tunnel studies domonstrate that up-wind buildings can have a significant effect on emissions from a chimney located within a few, say 5, building heights; for example, to maintain the same maximum ground level concentration the chimney height required in the presence of one building type studied would be between $1\frac{1}{2}$ and 2 times the height of the chimney required if the building was not present.

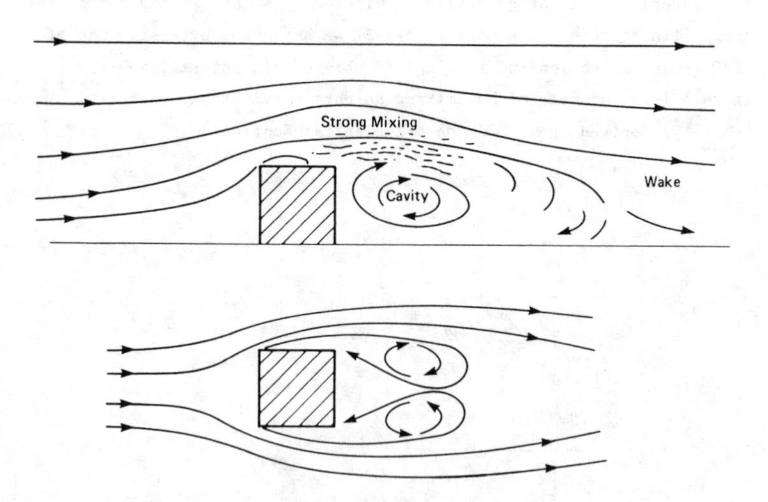

Figure 4 Simplified Schematic Flow Patterns Around a Cubical Building

Topographical features such as hills and sides of valleys can have similar effects on dispersion to those described above for buildings. Valleys are also somewhat more prone to problems arising from emissions fairly close to the ground. The incidence of low level or ground based temperature inversions can be greater, either because solar heating of the ground is somewhat delayed in the morning or because during the night cold air drains down the valley sides (katabatic winds) thus creating a "pool" of cold air on the valley floor. Any low level emissions will therefore disperse very slowly and may even accumulate to some extent. The relatively undiluted emissions can also drift along and across the valley thus affecting areas other than the immediate surroundings of the source. Emissions from high chimneys located on the valley floor may not be detected at all on the valley floor while the ground based inversion persists. However, considerable horizontal spreading of the plume aloft can occur and during the morning fumigation period large parts of the valley may experience relatively high concentrations at much the same time.

3. Dispersion Modelling

This lecture will be restricted to the application of the relatively simple, but proven, Gaussian-type model. This is not to condemn other more sophisticated models but for most practical applications the quality of the available emission and meteorological input data does not justify the much greater level of resources required to set up and run these other models. In many cases, a sound knowledge of meteorology and aerodynamics can be used to parameterise the Gaussian equation to simulate adequately, for example, the effects of buildings on dispersion.

3.1 Meteorological Data

Minimum requirements are wind speed and direction representative of the area in question plus estimates of stability categories calculated from local data or obtained from, for example, reference 5. Data on mixing heights are also of considerable value, if not essential, for the reasons given in 2.3.

3.2 Emission Data

The accuracy of data on emissions more than anything else determines the quality and accuracy of the modelled concentrations. Also, the spatial and temporal resolution of these concentrations can be no better than that for the emissions. With the assistance of Local Authorities and industrial companies, and using relevant Regional, etc. and land use statistics, fuel consumption inventories can be compiled[2,8,9] and augmented with information on industrial processes as appropriate. Relevant emission factors for each type of fuel and process can be used to estimate, for example, annual emission rates for all but the largest commercial and industrial premises for each $1km^2$ grid square. Emissions for the larger emitters can be calculated on an individual basis. In the East Strathclyde Study[6] there were some 1900 such sources and about 1200 grid squares. Figures 5 and 6 illustrate the spatial distribution of total emissions within the City of Glasgow, i.e. the central part of the inventory area. Table 3 is a disaggregation of the total emissions for the whole $1200km^2$ area by fuel and user type.

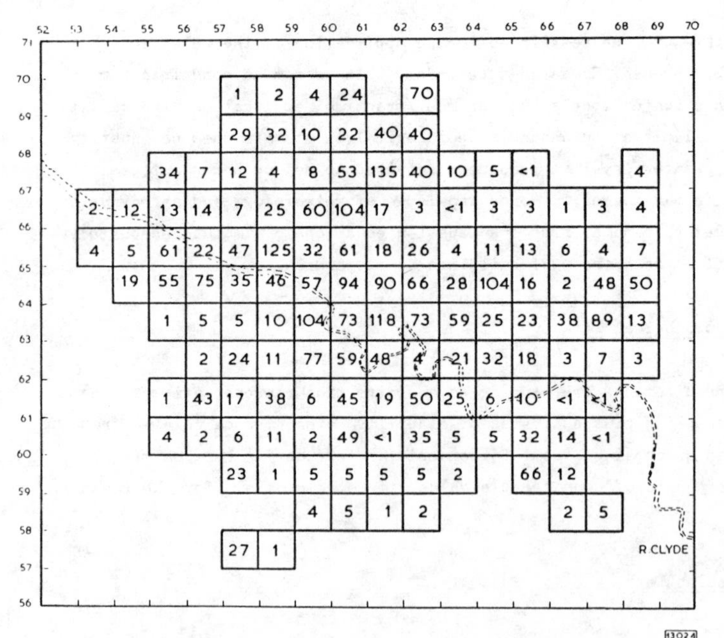

FIG. 5 TOTAL SMOKE EMISSIONS FOR THE GLASGOW AREA (tonnes/yr)

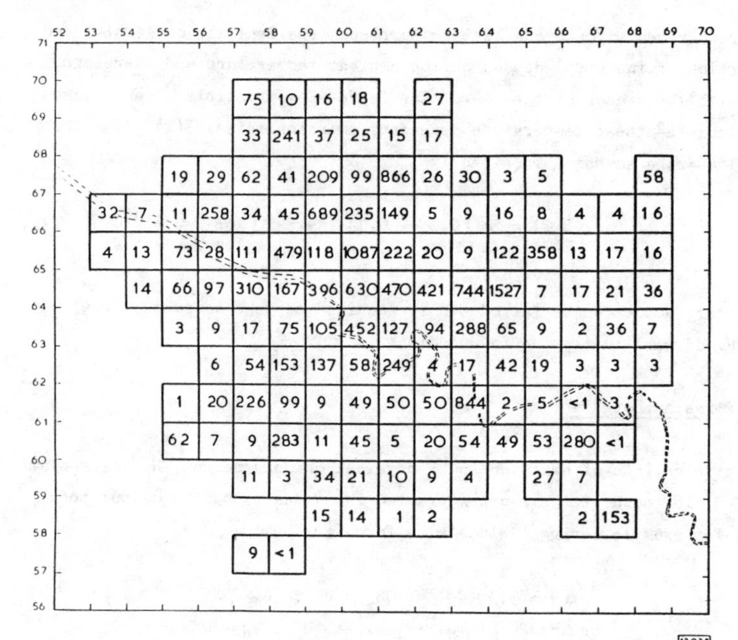

FIG. 6 TOTAL SO_2 EMISSIONS FOR THE GLASGOW AREA (tonnes/yr)

TABLE 3 - Estimated Emissions for the 1200 km^2 Inventory

Fuel	Emission Rate (tonnes/yr)					
	Indust./Commerc.		Domestic		Motor Traffic	
	Smoke	SO_2	Smoke	SO_2	Smoke	SO_2
Bituminous Coal	69	1040	6375	2035	-	-
Solid Smokeless	61	3406	367	754	-	-
Domestic Oil	-	-	1	284	-	-
Ind/Comm Oil	1556	46624	-	-	-	-
Motor Spirit	-	-	-	-	480	320
DERV	-	-	-	-	2726	909
Totals	1686	51070	6742	3073	3206	1229
Relative Emission (%)	14	92	58	6	28	2

Fuel requirements for space heating purposes (residential, commercial and a proportion of industry) depend on the ambient temperature and therefore vary with the season of the year. The Degree-Day principle[10] can be used to calculate[2] these temperature dependent emission rates, E(T), from the annual average emission rate (Eo):

$$\begin{aligned} E(T) &= Eo\,[0.33 + 0.11\,(14.5-T)] \quad \text{for } T < 14.5^{\circ}C \\ E(T) &= 0.33\,Eo \quad \text{for } T \geq 14.5^{\circ}C \end{aligned} \qquad (4)$$

A further factor can be introduced if required to take account of the typical diurnal variation in emissions.

3.3 Emission Heights

Effective heights of emission (He = physical chimney height, Hs, plus plume rise ΔH) for each of the sources treated individually can be calculated from, for example, Briggs' plume rise formula[11]:

$$\begin{aligned} \Delta H &= 3.3\,(Q_H)^{\frac{1}{3}}\,(10H_s)^{\frac{2}{3}}\,U^{-1}, \quad Q_H \geq 20MW \\ \text{or } \Delta H &= 20.5\,(Q_H)^{0.6}\,(H_s)^{0.4}\,U^{-1}, \quad Q_H < 20MW \end{aligned} \qquad (5)$$

where Q_H(MW) is the sensible heat emission from the chimney, and U is the wind speed (ms^{-1}) at height Hs (m).

For the sources treated collectively in grid squares, typical values of He can be used depending upon the average building height and boiler capacity for each category of emitter, e.g. 10-15 m for residential premises and 15-20m for small offices, etc.

3.4 Dispersion Equations

Equation (1) can be modified to take approximate account of the mixing height for a ground level concentration as follows:

$$X(x,y) = \frac{Q}{2\pi\sigma_y\sigma_z} \exp\left[-\frac{1}{2}\left(\frac{y}{\sigma_y}\right)^2\right]\left\{\exp\left(-\frac{He^2}{2\sigma_z^2}\right) + \exp\left(\frac{(2L-He)^2}{2\sigma_z^2}\right)\right\} \qquad (6)$$

where L(m) is the mixing height. This equation is not valid when He > L, during which circumstances the contribution from the emitter will often be zero.

Equation (6) is very sensitive to the values selected for the parameters σy and σz and considerable expertise is required in their selection and manipulation for all but the most simple situations. The parameters can conveniently be expressed in the form:

$$\sigma y = \sigma yo + a\ x^p \qquad (7)$$

$$\text{and } \sigma z = \sigma zo + b\ x^q \qquad (8)$$

Where x(m) is the distance downwind of the source, a,b,p and q are coefficients which are dependent on atmospheric stability and surface roughness and σyo and σzo can be used to take account of any initial spread of the plume, e.g. due to building entrainment.

Where long period averages are of concern the detailed dependence on σy is of much less importance and the pollutant concentrations can be assumed to be uniformly distributed cross-wind within each windsector. For a 30° sector the first two terms of equation (6) become

$$\frac{1{\cdot}524\ Q}{U\sigma z\ x} \qquad (9)$$

This applies to an individual chimney. For a grid square source x is replaced by (2·321d + x), where d(m) is the length of the side. For a receptor within a grid square a simple box model can be used to estimate the contribution from the sources within the square being treated collectively, viz

$$X = \frac{Q}{1{\cdot}244\ d\ U\ Hb} \qquad (10)$$

where Hb is the typical or average height to which the pollutants emitted from within the square are mixed whilst traversing the square. Values of Hb typically lie in the range 50 — 125 m, depending on the type of source and atmospheric stability.

One simple expression for σz, derived from Smith[4], is

$$\sigma z = \sigma zo + 0.9\ (0.83 - \log_{10} P).x^{0.73} \qquad (11)$$

where P is a parameter depending on atmospheric stability and lying in the range P=1 (A) to P=7 (G). For low level sources, where building entrainment is a feature of the dispersion, a typical value of σ_{zo} might be 10m. For higher level sources such as free standing industrial chimneys, σ_{zo} can usually be neglected.

3.5 The Modelling Process

All the items discussed in sections (3.1) to (3.4) together make up the dispersion modelling process. For a given receptor, the various equations are applied to each individual source or grid square as appropriate and for the relevant atmospheric stability, wind direction and wind speed. If long period averages are being calculated then the contributions for each combination of atmospheric stability, wind direction and wind speed must first be obtained and then weighted by the frequencies of occurrence of the different combinations.

If sources outside the area of direct concern are likely to make a significant contribution to pollutant concentration within the area then an estimate must be made of these "background" values, preferably by wind direction. In the Forth Valley Study[2] "background" contributions accounted for between 20 — 75% of the concentrations, depending on location, and there was a 3:1 variation in the "background" concentrations depending on wind direction.

3.6 Accuracy of Models

There are a variety of methods for validating or checking the accuracy of models although all essentially depend on comparison with observations obtained from monitoring networks. Ideally, these networks should be specifically designed and operated for that purpose. In making such comparisons it should be remembered that there are errors in measurements as well as in the modelling process. Furthermore, the model may not be able faithfully to reproduce the conditions around a given monitoring location because of the lack of spatial resolution in the emission inventory or for some other reason.

In general, it can be said that models perform better and more consistently in quantifying the proportional contributions from the different source categories than in quantifying the absolute concentrations. Thus, for example, if in a given area the model shows that residential sources are the major contributor to SO_2 concentrations then it can be assumed that the greatest improvement in air quality will be achieved by reducing residential emissions, even although there may be a much greater degree of uncertainty over the benefit in terms of absolute concentrations.

Two examples of comparisons between modelled and observed concentrations are given here. The first, in Table 4, compares the trend in spatial averages for the Forth Valley[2]. Overall, the agreement in absolute terms is reasonable and the year to year trend is fairly well reproduced. The reason for the modelled SO_2 values being consistently higher than the observed is thought to be due to the omission of dry and wet deposition in the model.

TABLE 4 - Modelled and Observed Spatial Averages

Winter	Smoke, $\mu g m^{-3}$		SO_2, $\mu g m^{-3}$	
	Modelled	Observed	Modelled	Observed
1973/4	29	30	48	42
1974/5	25	24	38	35
1975/6	26	27	39	32
1976/7	30	35	48	41

The second example is shown in figure 7, where the percentage error is defined as 100 x (modelled value - observed)/observed. The percentage of points refers to the fraction of the number of locations (N) for which direct comparisons between modelled and observed values were possible. For London, the modelled results were within ± 30% of the observed for 75% of the 20 receptors considered. This type of check on model accuracy should not be used if the number of receptors available for comparison is less than, say, 10.

Examples of Model Applications

Dispersion models are now in common use throughout the world. In addition to the two studies[2,8] already referred to, WSL has been making extensive use of modelling techniques for a variety of purposes including:

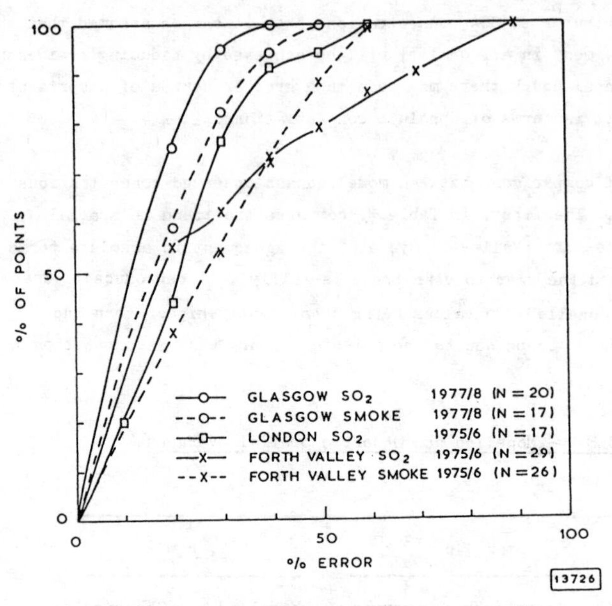

FIG. 7 PLOTS OF PERCENTAGE OF RECEPTOR POINTS WITHIN PERCENTAGES OF OBSERVED CONCENTRATIONS FOR ANNUAL AVERAGE SO_2 CONCENTRATIONS

i) Assessment of the implications of the EEC Smoke and SO_2 Directive[3] and identification of cost-effective controls.

ii) Prediction of the air quality impact of airport expansion at Heathrow, Gatwick and Stansted.

iii) Prediction of the impact of new power stations.

iv) Comparisons of the air quality impact of different options for locating petro-chemical developments.

Some of these applications have included pollutants other than smoke and SO_2, viz oxides of nitrogen, hydrocarbons, carbon monoxide and lead.

4. Acknowledgement

The author wishes to acknowledge that in compiling these lecture notes (this chapter) he has drawn extensively on published and unpublished work carried out by several colleagues within the Air Pollution Division at Warren Spring Laboratory.

5. References

1. Chimney Heights. Third Edition of the 1956 Clean Air Act Memorandum. London: HMSO, 1981.

2. Keddie A.W.C. et al. The Measurement, Assessment and Prediction of Air Pollution in the Forth Valley of Scotland - Final Report . Warren Spring Laboratory Report LR 279(AP), Stevenage, 1978.

3. Official Journal of the European Communities, 30 August 1980, L229, 30-48, Directive No. 80/779/EEC.

4. Pasquill, F. Atmospheric Diffusion. Chichester: Ellis Horwood, 1974.

5. A Model for Short and Medium Range Dispersion of Readionuclides Released to the Atmosphere. First Report of a UK Working Party on Atmospheric Dispersion, Editor R.H. Clark. London: HMSO, 1979, NRPB Report R91.

6. R.A. Maughan, Frequency of potential contributions by major sources to ground level concentrations of SO_2 in the Forth Valley, Scotland. Atm. Environ., 13, 1697-1706, 1979.

7. R.A. Maughan, A.M. Spanton and M.L. Williams, An analysis of the frequency distribution of SODAR derived mixing heights classified by atmospheric stability. Atm. Environ., 16, 1209-1218, 1982

8. Williams, M.L. et al. The Measurement and Prediction of Smoke and Sulphur Dioxide in the East Strathclyde Region: Final Report. Warren Spring Laboratory Report LR 412(AP), Stevenage, 1981.

9. D.J. Ball and S.W. Radcliffe, An Inventory of SO Emissions to London's Air, GLC Research Report No. 23, London, 1979.

10. Degree Days. London: Gas Council, Technical Handbook No. 101.

11. Briggs G.A., Plume Rise, US Atomic Energy Commission, 1969.

11
Legislation and the Control of Air Pollution

By. F. E. Ireland
CONSULTANT IN AIR POLLUTION CONTROL, 59 LANCHESTER ROAD, HIGHGATE, LONDON N6, U.K.

Introduction

Early legislation in environmental control is lost in the mists of time. During the reign of Julius Caesar a law was passed which stated that "Currus Noctu Vitenur" - let chariot driving be banned at night - presumably for the disturbance it caused. Then we are told of the man in London in the 14th Century who was hanged for causing a nuisance to his neighbours by burning coal. For centuries Common Law protected individuals in the UK from specific nuisances due to polluted air. In their book "The Politics of Clean Air", Lord Ashby and Mary Anderson refer to the case in 1691 when Thomas Legg of Coleman Street, London, complained of the smoke from his neighbour's bakehouse; the baker was ordered to put up a chimney "soe high as to convey the smoake clear of the topps of the houses". But Common Law is made to protect people, not air. If air is to be protected, whether or not it is causing a nuisance, the State has to intervene. However, it has not always been easy to have legislation passed in Parliament on pollution control, even when the damage was obvious and many people were being killed by it, especially during the London smogs. As an example let us take the Clean Air Act 1956, when the circumstances leading to the Act were the London smog of 1952, the increased death rate attributable to the smog and the attempt by Gerald Nabarro, MP, to introduce a private Member's Bill. The historical roots of the Act go well back into the last century, probably to 1819 when Parliament appointed a select committee to enquire 'how far it may be practicable to compel Persons using Steam Engines and Furnaces in their different Works to erect them in a Manner less prejudicial to public Health and public Comfort'. Little came of the inquiry beyond a slight stiffening of the Common Law on nuisance. Twenty years later there began a sustained campaign to suppress smoke, which has gone on ever since, and it provided rich material for study, by Ashby and Anderson, of the beginnings of social response to what they call 'the unacceptable face of technology'. One of the most important aspects was the awakening of public opinion to the evils of air pollution and the pressures on Parliament to have something done to control it. The technology for air pollution control has been known for a long time and engineers are well aware of it and of the consequences of not controlling the hazards involved. As one group has put it, 'The finest technical solution to a problem is worthless if it cannot be explained or sold to the public and consummated'. In dealing with environmental problems it is not enough to apply hard data, to quantify the issues, and to expect a rational response. Logic penetrates the head, but not the heart, and many of these issues are settled by the heart, especially in Parliament where the final decisions are political. Very often the economists will try to put values on imponderables for what is efficient for society. But the public wants to know what is good for society, and the most precious kinds of good cannot be quantified without distortion.

So, in order to have legislation passed in Parliament today environmental education is essential in a number of ways. The engineers and scientists in their technical education must be made aware of the environmental consequences of their work; they must be taught some understanding of the management of conflict; and they must be able to accept public participation. The environmental engineer has to know how to incorporate unquantifiable values into the decision equations for environmental policies; and also how to reconcile the technologist's judgment about an environmental issue based on hard data with popular judgment, which may have no rational relationship to the hard data. Conflicts end with decisions, usually by politicians, and the politicians, mindful that there are more votes in emotion than in logic, are tempted to trade long-term benefits for short-term approval from their constituents. The most critical outcome of such decisions may be a serious weakening of confidence in the institutions of government. Engineers with an appreciation of the environmental consequences of their work can make a great contribution to the process of social innovation, especially in the interpretation of highly technical issues.

Legislation about air pollution control has always been emotive, and still is, as evidenced by the present controversy over lead in petrol. The government set up an expert working party to report on the medical effects of lead in the environment and to make recommendations. This was the most thorough investigation on the subject ever made in the UK and the report was published in March 1980. On the day the report was issued, the anti-lead lobby denounced this authoritative scientific study as a 'cover-up' and a 'political document'. The scientists were labelled 'establishment figures' who had succumbed to pressure from the oil lobby. These critics rejected the evidence because it did not support the popular prejudice, which was itself encouraged by the media.

Clean Air Legislation in the UK

Planning

In the United Kingdom, air pollution control is exercised by both local and central governments. Control actually begins with Land Use by means of the Planning Acts and if the planners do their job properly there should be fewer conflicts between domestic, commercial, industrial, agricultural and aesthetic interests than there have been in the past. Too often have we seen clean urban occupations, eg. schools, having to operate alongside other occupations producing noisy, dirty or obnoxious emissions. Applications for the erection of new structures, or for new activities, have to be submitted to the Planning Authorities who decide whether or not to grant planning permission. It is now a modern requirement that in industrial cases environmental impact statements have to be submitted with the applications and the contents of the statements vary with the complexity of the cases. Impact statements have become an important part of planning applications and can be very costly to prepare, especially where large, controversial schemes are proposed. There is, of course, right of appeal to the Secretary of State over the resulting decision, but in certain instances the Secretary of State may call in the case, or have it referred to him, when he must hold a public inquiry or receive evidence from interested parties. He then takes his own decision after considering the recommendation of the chairman of the inquiry. Both the planning authority and the Secretary of State can put conditions on the consent, and then those responsible for implementing the controlling Acts take charge of emissions to air from the premises when commissioned.

Consents for the use of premises are only given for nominated types of use described under the Use Classes Order, and the use cannot be changed from one class to another without the permission of the planning authority. This is an essential precaution to prevent an owner changing from a clean to a dirty process in a neighbourhood where it would interfere greatly with the amenities.

Transport, storage, noise, air and water pollution, residue disposal and aesthetics are some of the important local considerations for planning authorities to consider when examining applications. More attention is being given these days to the appearance of new plant and buildings within the bounds of reasonable expenditure, and the scope for fitting buildings pleasantly into the surroundings. Particular care is being given to chimney heights and landscaping, and there is sometimes a conflict of interests between the planners and environmental health officers of the same authority on the score of chimney heights being too high for the one and too low for the other.

Division of Responsibility

Local authorities exercise air pollution control chiefly through the Public Health, Clean Air and Control of Pollution Acts, and the central Authority through the Alkali &c Works Regulation Act 1906 and the Health and Safety at Work Etc Act 1974.

To get matters in perspective, there are some 250,000/300,000 commercial and industrial enterprises in the UK and it is estimated that about 30,000/50,000 have emissions to air which need the special attention of local or central authorities in order properly to control them. Of these 30,000/50,000 works, about 2,000 are scheduled under the Alkali Act because they need the special attention of a professional inspectorate with expertise in chemistry, chemical engineering and fuel technology in order that the physico-chemical processes concerned can be properly understood and controlled.

Local authorities generally implement their controlling Acts by means of their Environmental Health Officers, who have many other important duties to perform, although some of the larger towns and cities have separate air pollution control teams.

Local Authority Control

Public Health Act 1936 and Public Health(Recurring Nuisances) Act 1969

The nuisance provisions of Sections 91 to 100 of the 1936 Act are used to control emissions and accumulations which are prejudicial to health or a nuisance, and which are not covered more specifically by the provisions of other Acts, such as the Clean Air Acts. Proving a nuisance is often difficult and the Courts have not always been as helpful as they might have been in punishing offenders, but the 1936 Act does not operate until a nuisance occurs. This was remedied to some extent by the Public Health (Recurring Nuisances) Act 1969 which gave local authorities powers to anticipate nuisances and to take action beforehand, especially when there was a danger of a nuisance recurring. The Public Health Acts are useful in acting as a control over miscellaneous emissions, eg. dust and odours, which do not come under the Clean Air or Alkali Acts.

Clean Air Acts 1956 and 1968

The 1956 Act is the principal Act and its scope and powers were extended by the 1968 Act. These Acts control emissions to air of smoke, grit, dust and fume from combustion processes, and allow local authorities to decide on the heights of new chimneys and chimneys for extensions to existing furnaces and to set up smoke control areas. Briefly, requirements are as follows:

Smoke.- Dark smoke, that is smoke as dark as or darker than shade 2 on the Ringelmann Chart, shall not be emitted from chimneys, or from the burning of combustible matter on the ground. Local authorities have powers to designate their areas, or parts of their areas, as smoke control areas, wherein it is an offence to emit smoke and only authorised fuels, or classes of appliances exempted by the Minister because he is satisfied that they can burn fuel other than authorised fuels without producing smoke or a substantial quantity of smoke, can be used. In order to identify solid fuels which could be regarded as authorised fuels, the British Standards Institution issued BS.3841:1965, "Method for the Measurement of Smoke from Manufactured Solid Fuels for Domestic Open Fires". It also recommended an emission standard which was accepted by the Minister, that is, when manufactured solid smokeless fuels are tested in the manner prescribed by BS.3841:1965 the rate of smoke emission shallnot be greater than 5 grammes per hour. This represents an emission of not more than 20 per cent of the smoke emitted on average when bituminous coal is burnt under similar test conditions, and is in line with the 'Beaver' Committee's recommendation that emissions of smoke should be reduced by at least 80 per cent. Gas and electricity are included in the list of authorised fuels, but oil is not, although certain oil-burning appliances have been approved. The Minister has exempted certain processes from the provisions of the Clean Air Acts where it is known that there are technical difficulties in meeting the requirements, including those processes scheduled under the Alkali Act, when best practicable means must be used to prevent emissions to air.

Grit, Dust and Fume. - The Acts require that any practicable means must be used to minimise the emission of grit, dust and fume, and that new furnaces burning (a) pulverised fuel; or (b) solid matter at a rate of 100 pounds or more per hour; or (c) any liquid or gaseous matter at a rate equivalent to $1\frac{1}{4}$ million or more British Thermal Units an hour, must be fitted with plant for arresting grit and dust, which has been approved by the local authority. The Minister may exempt prescribed classes of furnace from these provisions, and local authorities may exempt individual furnaces if they are satisfied that emissions will not be prejudicial to health or a nuisance. So far, the Minister has issued two regulations in these categories. Statutory

Instrument 1971 No.161, The Clean Air(Measurement of Grit and Dust from Furnaces) Regulations 1971 specifies the adaptations to chimneys for making measurements of emissions and the provision and maintenance of apparatus by one of the procedures described in British Standard 3405, 1961(later replaced by the 1971 version). Statutory Instrument 1971 No.162, The Clean Air(Emission of Grit and Dust from Furnaces) Regulations 1971 sets limits on the rates of emission of grit and dust from furnaces in which the material being heated does not contribute to the emission, eg. boilers, metal reheating furnaces. There is also a limit on the proportion of grit, ie. particles exceeding 76 microns in diameter, which may be emitted in the total particulates; 33% where the maximum continuous rating does not exceed 16,800 pounds per hour of steam or 16,800,000 BTUs per hour, or 20% in any other case.

Working parties set up by the Department of the Environment have made recommendations for other cases of emission of grit and dust where the material being heated contributes to the emission, but the Minister has not yet issued regulations because of the present state of the economy.

Chimneys. - The owner of a furnace burning solid matter at a rate of 100 pounds or more per hour, pulverised fuel, or liquid or gaseous matter at a rate equivalent to $1\frac{1}{4}$ million BTUs or more per hour, shall not use the furnace unless the height of the chimney serving the furnace has been approved by the local authority, who must be satisfied that the height will be sufficient to prevent, so far as practicable, the smoke, grit, dust, gases or fumes from being prejudicial to health or a nuisance having regard to

(a) the purpose of the chimney;
(b) the position and descriptions of buildings near it;
(c) the levels of the neighbouring ground;
(d) any other matters requiring consideration in the circumstances.

This section applies to new furnaces, to any furnace the combustion space of which is increased, or which replaces a furnace which had a smaller combustion space. There is a right of appeal to the Minister on local authority decisions.

In order to assist local authorities and industrialists in the determination of rational chimney heights in a co-ordinated fashion throughout the UK, the Minister in 1963 issued the Memorandum on Chimney Heights, replaced in 1967 by the booklet "Chimney Heights", which was an extension of the original. The booklet set out to provide a relatively simple method of calculating the approximate height commonly desirable in normal circumstances. It is a guide rather than a mathematically precise way of reaching a final decision on chimney height. The method of calculation is based on the amount of flue gases which the chimney is expected to emit as a function of the maximum rate of emission of sulphur dioxide. The character of the surrounding area is taken into account before reference is made to a series of nomograms giving the 'uncorrected chimney height'. If the uncorrected chimney height is greater than $2\frac{1}{2}$ times the height of nearby buildings, it becomes the final chimney height. Corrections are necessary when the uncorrected height is less than $2\frac{1}{2}$ times the height of such buildings and these establish the final chimney height by means of a second set of nomograms based on the formula $H = 0.56A + 0.375B + 0.625C$.

where H = final chimney height, feet.
A = building height or greatest length, whichever is the lesser, feet.
B = building height, feet.
C = uncorrected chimney height, feet.

The provisions of the Clean Air Acts mentioned above do not apply to any work subject or potentially subject to the Alkali &c. Works Regulation Act 1906, where best practicable means must be used to render emissions harmless and inoffensive.

The Control of Pollution Act 1974

Part 1V of the Control of Pollution Act 1974 is concerned with Pollution of the Atmosphere and is divided into two aspects dealing with prevention of atmospheric pollution and information about atmospheric pollution.

Prevention of atmospheric pollution. - This aspect is mainly concerned with the Secretary of State's powers to issue regulations imposing requirements as to the composition and content of oil fuel which is used in furnaces or engines. It has enabled the Secretary of State to implement an EEC Directive by issuing a regulation limiting the sulphur content of gas oil to 0.3 per cent and the government has announced its intention to reduce the maximum amount of lead in petrol to 0.15 g/l by 1985.
There is a small section clarifying the position of cable burning under the Alkali Act.

Information about atmospheric pollution. - This is a revolutionary step in the UK air pollution control policy. Following the recommendations of a working party, the Act has given powers to local authorities to obtain information about the emission of pollutants and other substances into the air either by measuring and recording the emissions themselves from premises (not being private dwellings), or by entering into an arrangement with occupiers under which the latter measure and record emissions on behalf of the local authority. Notices have to be given to the occupier by the authority. The information so collected shall be for the purpose of publication in a register to be kept by the authority, who shall consult local industry and other interests not less frequently than twice a year about the way in which the information collected shall be made available to the public in a meaningful form and without giving away trade secrets. The Secretary of State shall by regulations prescribe the manner in which, and the methods by which, local authorities are to perform these functions, and in particular require returns of:

(a) the total volume of gases, whether pollutant or not, discharged from the premises in question over any period;
(b) the concentration of pollutant in the gases discharged;
(c) the total of pollutant discharged over any period;
(d) the height or heights at which discharges take place;
(e) the hours during which discharges take place;
(f) the concentration of pollutants at ground level.

Notices shall not require returns at intervals of less than three months and no one notice shall call for information concerning a period of more than twelve months.

Central Authority Control

Alkali &c. Works Regulation Act 1906 and Health and Safety at Work Etc. Act 1974

These Acts are taken together because the Alkali Inspectorate was merged into the Health and Safety Executive on 1 January 1975 and some of the provisions of the Alkali Act have been replaced by regulations under the Health and Safety at Work Etc. Act 1974. It is the ultimate intention to replace all of the Alkali Act provisions by regulations under the new 1974 Act.

The first Alkali Act was passed in 1863 to control emissions of hydrochloric acid gas from the first stage of the Leblanc process for production of alkali, or sodium carbonate. Since the first Act, many other processes have been scheduled and today 62 types of works are registrable. These include much of the chemical industry, oil refineries, petrochemical processes, electricity generation, coal carbonising, iron and steel production, non-ferrous metals, cement, lime, ceramics, and the like, where a knowledge of chemistry, chemical engineering and fuel technology are needed in order properly to understand and control the physico-chemical processes giving

rise to emissions. The Acts are administered by inspectors of the central authority. Full lists of the registrable works and noxious or offensive substances are given in Reference 1. Section 3 of the Health and Safety at Work Etc. Act 1974 also puts a general duty on employers to ensure that persons not at work are not exposed to risk to their health and safety by reason of the employers' activities. The basic requirements of the Alkali Act are that scheduled works must use the'best practicable means' (BPM) (a) to prevent the emission of noxious or offensive substances, and (b) to render harmless and inoffensive such substances as are necessarily discharged. The expression 'best practicable means' has reference not only to the correct use and effective maintenance of equipment installed for the purpose of preventing emissions to air, but also to the proper control, by the owner, of the process giving rise to the emission. The word "practicable" has regard for the local circumstances, the financial implications and the current state of technical knowledge. The term "financial implications" is treated by the inspectorate as having a much wider meaning than just the cost to the owner, and account is taken of the effect on local employment, rates, local services and other people who may be dependent on the wage earners if too tough a policy causes redundancies or even works closure.
Only in four cases does the Alkali Act specify statutory standards of emission and in order to help industry and the inspectorate to judge what constitutes BPM the chief inspectors have set their own presumptive standards and other requirements, which they can alter at will to take account of improving technology and the demands of the public for a better environment. In practice the chief inspector consults with the industries concerned and other interested parties before making his decisions. These Notes on Best Practicable Means are published by the Health and Safety Executive for a number of scheduled processes and an example for Petroleum Works(PVC polymer plants) is given in Reference 2.

The setting and maintaining of standards of emission from chimneys is only a part of BPM. There are many sources of airborne pollution from industry for which standards cannot be set. The handling, storage and transport of dusty materials, spillages, accidents, leakages from ill-maintained plant, flaring of waste gases, discharge of materials from process vessels, poor housekeeping, are but a few examples. In such cases, codes of good practice are written and the experience of the inspectorate is used in inspecting and enforcing the most practicable means of abatement, maintenance, enclosure and supervision. The presumptive standards and other requirements of the inspectorate have legal force, because works not meeting them are presumed to be guilty of not using BPM. The ultimate sanction is prosecution and Section 33 of the Health and Safety at Work Etc. Act 1974 lays down penalties for offences as follows:

(a) on summary conviction, to a fine not exceeding £1000;

(b) on conviction on indictment - (i) if the offence is one to which this sub-paragraph applies, to imprisonment for a term not exceeding two years, or a fine, or both; (ii) if the offence is not one tb which the preceding sub-paragraph applies, to a fine.

The offences to which the above penalties apply are detailed in Section 33 of the Act.

The Health and Safety at Work Etc. Act also gives inspectors the powers to issue Improvement or Prohibition notices. The first requires the owner to remedy the contravention within a stated time, whilst the second type takes immediate effect.

The second part of best practicable means as stated earlier, requires the use of BPM to render harmless and inoffensive such substances as are necessarily discharged. This is achieved by dispersion from suitably tall chimneys, but these are only considered after the BPM for prevention have been fully explored. For small sources of waste gases from the burning of fuel, the inspectorate uses the booklet "Chimney Heights" referred to earlier, but large emitters, such as power stations, receive special mathematical consideration, generally using the Sutton-Bosanquet formulae for dispersion and plume rise due to buoyancy, basing the calculation on sulphur dioxide emissions. However, industry emits many other pollutants than sulphur dioxide, some of a highly toxic nature, and the inspectorate has formulated rules for determining chimney heights for such substances, taking into account many factors such as the ground-level concentration under normal and accident conditions, short- and long-term effects, the targets which may be affected, etc. These are described by Ireland in Appendix V of the 106th Annual Report on Alkali &c. Works 1969. They are constantly under review as more exact information on medical effects, dispersion, plume rise and atmospheric reactions becomes known. The aim is to restrict the contribution which any one source may make to the environment to a safe level and to permit the operation of multiple sources within the same locality, whilst still maintaining an acceptable air quality.

International Legislation

The European Communities' pollution control policy is clearly defined by a collection of instruments which in one form or another have been concerned, approved, or adopted by the Community institutions. At the general level, the European Communities' policy on pollution control is contained in the Declaration of the Council of the European Communities and of the

Representatives of the Governments of the Member States meeting in the Council of 22 November 1973, on the programme of action of the European Communities on the Environment, and in the Resolution of the Council of the European Communities and of the Representatives of the Governments of the Member States meeting within the Council of 17 May 1977 on the continuation and implementation of a European Community policy and action programme on the environment.

In addition to these two basic documents, there are a number of specific instruments which have been adopted by Council and give more details. There are also proposals for Community action which have been approved by the Commission but have not yet been approved by Council.

The action programme called for a series of projects to be undertaken at Community level in order to provide a common basis for the evaluation of data and a common framework of methods and references. Amongst the tasks which would have to be undertaken was the laying down of scientific criteria for the degree of harm of the principal forms of air and water pollution and for noise, priority being given to lead and lead compounds, organic halogen compounds, sulphur compounds and particles in suspension, nitrogen oxides, carbon monoxide, mercury, phenols and hydrocarbons.

On 15 July 1980, the European Community issued its Council Directive on air quality limit values and guide values for sulphur dioxide and suspended particulates, and these are shown summarised in Tables 1 and 2 below.

Table 1

Limit Values for Smoke and Sulphur Dioxide in microgrammes per cubic metre.

Period	Smoke	Limit values for SO_2	
Year(median of daily values)	80 (68)	If smoke less than 40:	120
		If smoke more than 40:	80
Winter(median of daily values Oct.-March)	130 (111)	If smoke less than 60:	180
		If smoke more than 60:	130
Year(Peak) (98 percentile of daily values)	250 (213)	If smoke less than 150:	350
		If smoke more than 150:	250

Limit values for smoke relate to the OECD method of measurement: figures in brackets give equivalents for the BSI method as used in the National Survey.

Table 2

Guide Values for Smoke and Sulphur Dioxide in microgrammes per cubic metre.

Period	Smoke	Sulphur Dioxide
Year(arithmetic mean of daily values)	40 to 60 (34 to 51)	40 to 60
24 hours. (daily mean value)	100 to 150 (85 to 128)	100 to 150

BSI values in brackets.

Air Quality standards are a logical way of expressing the ultimate objective of clean air, but they do not of themselves provide a simple guide to the control policy that must be adopted to meet them. Implementation in the UK is provided by the Clean Air and Alkali Acts, but the EEC must ensure that implementation throughout the Member States is uniform if distortion of commercial interests is to be avoided. The UK has a far better record of inspection and enforcement than most other countries.

In order to reduce sulphur dioxide concentrations within large urban areas, which stemmed from numerous low level sources, the Council issued a Directive on 24 November 1975 limiting the sulphur contents of the only two grades of gas oil to be marketted. The present requirement, from 1 October 1980, is that the maximum sulphur content of type A shall be 0.3% and that of type B 0.5%.

The Commission has made proposals to the Council for limits to be placed on the sulphur content of residual fuel oil, but these are still under discussion. It is a very difficult subject with many ramifications.

Pollution by Lead. On 27 March 1977 the Council adopted a Directive on biological screening of the population for lead. Sampling of blood lead was to be carried out over a four-year period by Member States and this has been implemented in the UK. In assessing the results of the biological screening the following blood lead levels should be taken, together, as reference levels.

- a maximum of 20µg of Pb/100 ml of blood for 20% of the group of people examined.
- a maximum of 30µg of Pb/100 ml of blood for 90% of the group of people examined.
- a maximum of 35µg of Pb/100 ml of blood for 98% of the group of people examined.

The Commission also proposed another Directive for air quality standards for lead where there would be no specific effect on the lungs.

(a) an annual mean level of not more than 2μgPb/m^3 in urban residential areas and areas exposed to sources of lead other than from traffic;

(b) a monthly median level of not more than 8μg Pb/m^3 in areas particularly exposed to motor vehicle traffic.

The Commissions proposals are still being considered by the Environment Working Group of the Council.

On 30 May 1978 the EEC environment Ministers adopted a Directive which provided that from 1 January 1981 the maximum permitted lead compound content in petrol, calculated as lead, should be 0.4 g/l; Member States should not establish limits lower than 0.15 g/l. The UK has implemented this Directive and the government has given notice that in 1985 the lead content of petrol sold in the UK shall not exceed 0.15 g/l.

It seems appropriate to mention here the recent development by a UK company of lead-tolerant catalysts for motor vehicle exhausts to reduce emissions of carbon monoxide, hydrocarbons and oxides of nitrogen. Earlier catalysts for this duty are poisoned by lead in the petrol and this is the reason why some countries have specified lead-free petrol. In the presence of strong sunlight the pollutants from vehicle exhausts can react in the atmosphere to form the notorious photochemical smog. Motor vehicle exhausts are dirty in their own right and even if the UK does not suffer significantly from photochemical smog there is a strong case for having them cleaned, because they contribute a major proportion of the existing pollution in UK urban areas. The reduction of lead in petrol to 0.15 g/l will make easier the operation of lead-tolerant catalysts to clean up vehicle exhausts and it is to be hoped that the UK and the EEC will issue legislation on this subject.

On 28 May 1974 the EEC issued a Directive (revising a 1970 Directive) limiting the mass rates of emission of carbon monoxide and hydrocarbons by motor vehicles, but in the author's view another look should be given in the light of technical developments. This EEC Directive was based on the technical requirements elaborated by the UN Economic Commission for Europe (ECE). The Council is considering Directives for limiting emissions of oxides of nitrogen from vehicles and of smoke from diesel-engined vehicles.

The Future

In the future we can expect an increasing influence by the EEC on the National legilation of Member States in the field of air pollution control. The Council has, inter alia, called on the Commission to study pollution problems arising in certain industrial sectors and in energy productions, giving priority to industries emitting dust, oxides of sulphur and nitrogen, hydrocarbons and solvents, fluorides and heavy metals.

In the UK, it has been recognised for a long time that there is a gap in local authority control over air pollution from non-combustion sources. Reliance is placed upon the nuisance provisions of the Public Health Act 1936, but government has appreciated a need for a more positive approach by giving powers at an appropriate time for prior approval of industrial plant and regular inspection by local government officers. In other words, a BPM approach.

Amongst other things, the Fifth Report of the Royal Commission on Environmental Pollution recommended the setting up of a body known as Her Majesty's Pollution Inspectorate, based on the Alkali and Clean Air Inspectorate. This new body would be responsible for controlling the generation of liquid and solid wastes from industry in the same way as the Alkali Inspectorate does at present for air pollution control. The setting of standards and the disposal of the resulting liquid and solid wastes would remain with the existing bodies. At present there is only responsibility for disposal and not for generation. The government has not yet responded to this recommendation.

References

[1] 'Lists of registrable works and noxious or offensive gases specified in the Act and Orders' (Alkali &c. Works Regulation Act 1906; Alkali &c. Works Orders 1966 and 1971), Health and Safety Executive, London.

[2] 'Notes on Best Practicable Means: Petroleum works (PVC polymer plants)', Health and Safety Executive, London.

12
Catalyst Systems for Emission Control from Motor Vehicles

By G. J. K. Acres
JOHNSON MATTHEY RESEARCH CENTRE, BLOUNT'S COURT, SONNING COMMON, READING, BERKS. RG4 9NH, U.K.

INTRODUCTION

The possible harmful effects of gaseous emissions from motor vehicles and the use of catalyst systems for their control were first reported in patents and publications dating from 1925. Not until 1975 was the need to use catalysts established and the technology sufficiently advanced for this demanding application. Catalyst systems are now used on the majority of cars sold in America and Japan, while proposed emission control legislation in Europe could necessitate the use of catalysts before the end of the decade.

EXHAUST EMISSIONS FROM MOTOR VEHICLES

The principal gaseous emissions from petrol engined vehicles are carbon monoxide, unburnt hydrocarbons and nitrogen oxides. Depending upon the density of vehicles in a particular area, the motor vehicle can be responsible for up to 50% of hydrocarbon and nitrogen oxide and 90% of carbon monoxide emissions to the atmosphere. While having some direct environmental effects on health, the major problem arising from these emissions is photochemical smog. This phenomenon, first observed in 1940 in Los Angeles, results from the reaction of hydrocarbons, nitrogen oxides and oxygen initiated by sunlight.

While gaseous emissions from motor vehicles can be significantly reduced by engine design and modification, the levels required by current American and Japanese legislation have not yet been achieved by such changes without substantially impairing driveability and fuel economy of the vehicle. Thus the major benefit of using catalyst systems for emission control is to allow the engine to be optimised for these important factors.

CATALYST SYSTEMS

Since carbon monoxide and hydrocarbon emissions are removed by oxidation whereas nitrogen oxides may only be removed by reaction with reducing agents, the elimination of all three pollutants can only be achieved by separating the oxidation and reduction functions in appropriate reactors or by design of a selective catalyst on which the two reactions may proceed simultaneously. The presentation presented the development and use of oxidation, dual bed and three way catalyst systems with particular reference to catalyst durability using slides and film to illustrate important physical parameters as described in References 1 and 2.

Lead Tolerant Catalysts

The development of the catalyst systems now widely used on cars sold in America and Japan was made somewhat easier by the decision of these countries to phase-out the use of lead additives in gasoline by the introduction in 1975 of lead-free fuel. The durability of these catalysts, which have a life in excess of 50,000 miles, would be seriously impaired if used on cars using leaded fuel.

In some countries, notably Europe, there is currently a preference to retain at least some lead in the fuel so as to retain high compression engines with their fuel economy benefits. The presentation outlined recent developments in lead tolerant catalyst technology aimed at future European requirements and is described in Reference 3.

DIESEL ENGINE EMISSION CONTROL

Notably in Europe, but increasingly in America, diesel engined vehicles are widely used where reliability and fuel economy in city driving are important factors. While this engine has relatively low carbon monoxide and hydrocarbon emissions, recent work has shown that the high particulate emissions can contain polynuclear aromatic fractions which may constitute an environmental problem. The presentation described the development of a catalyst system designed for use with diesel engined vehicles. The information presented is given in Reference 4.

THE CATALYTIC ENGINE

Catalyst systems for the treatment of exhaust gases are now widely used or are in an advanced stage of development for both petrol and diesel engined vehicles. However, a substantial effort has been and continues to be made to develop a power system for vehicles which combines the benefits of low emissions and good fuel economy with the ability to operate on a wide range of fuels without significantly increasing the cost of the system. By using a catalyst to initiate combustion in the combustion chamber, in principal, the present limiting factors in the design of both petrol and diesel engines could be eliminated thus providing some or all of the benefits outlined above. In the presentation the development of an engine using catalytic ignition was described and is given in Reference 5.

In conclusion therefore, catalyst technology now plays a major part in controlling emissions from motor vehicles and further developments highlighted in the presentation are likely to lead to its more widespread application in this decade.

REFERENCES

1. G.J.K. Acres and B.J. Cooper, Platinum Metals Review, 1972, 16, 74.
2. B. Harrison, B.J. Cooper and A.J.J. Wilkins, Platinum Metals Review, 1981, 25, 14.
3. A.F. Diwell and B. Harrison, Platinum Metals Review, 1981, 25, 142.
4. B.E. Enga, Platinum Metals Review, 1982, 26, 50.
5. R.H. Thring, Platinum Metals Review, 1980, 24, 1.

13
Evaluating Pollution Effects on Plant Productivity: A Cautionary Tale

By T. A. Mansfield
DEPARTMENT OF BIOLOGICAL SCIENCES, UNIVERSITY OF LANCASTER, LANCASTER LA1 4YQ, U.K.

When, early in the present century, people began to evaluate the effects of air pollutants on plants they were unaware of the difficulties that lay ahead. Realisation of the extent of these difficulties has developed only slowly, and much of the literature that has accumulated is so confusing that there are few clear examples of the kinds of information that a chemist may seek, for example dose-response relationships and toxicity thresholds. We are, however, making progress insofar as we are beginning to realise the complexity of the situation with which we are dealing. In 1949, M. Katz[1], a worker of long experience, included this statement in a review written primarily for industrial chemists: "........it is evident that sulfur dioxide concentrations up to approximately 0.30 ppm may be present in the atmosphere of agricultural areas during the growing season and not prove detrimental to vegetation ...." If I were called upon to amend this statement in the light of studies over the last 30 years I would replace 0.30 ppm by 0.03 ppm, but would be profoundly uncomfortable about giving the impression that I had hard facts at my disposal. Real advances are, however, being made now that we understand that many external and internal factors can affect a plant's reaction to pollutants. The present situation is aptly summarised by some words of Alphonse Karr: "The more things change, the more they are the same" (Les Guêpes, 1849).

The pollutant whose effects on plant productivity have been most thoroughly investigated is SO_2. I have decided to spend most of this lecture discussing our present understanding of these effects, because the historical development of experimental studies with SO_2 provides the best illustrations of the difficulties encountered by plant scientists in providing adequate estimates of responses to pollutants in the field.

<u>Visible and invisible injury</u>

There is often visible damage to vegetation around industrial sources of SO_2, and also in the vicinity of natural ones such as volcanoes. The early work on the subject was stimulated by observations of such injury and eventually the idea developed that compensation for lost agricultural production might be based on the percentage of the leaf area visibly damaged. Acute injury due to SO_2 is the result of the collapse of cells in leaf tissues, and the formation of distinctive necrotic areas. Since the primary process behind crop production is photosynthesis, and since the interception of light for photosynthesis is directly proportional to leaf area, the simplistic view was taken that a reduction in the area of green leaves should be directly related to crop losses. Suggestions that there might be invisible injury, namely a reduced photosynthetic efficiency of leaves not visibly damaged, were strongly disputed when they were made by workers such as Stoklasa[2]. Reviewing the subject in 1949 and 1951, respectively, M. Katz[1] and M.D. Thomas[3] were in no doubt that the invisible injury theory was without foundation. Katz expressed the hope that it had "....been disposed of, once and for all time...." and Thomas echoed his sentiments, writing "......a large amount of experimental work has been done which demonstrates that this theory is entirely without foundation".

The influence of these two senior scientists was so great, particularly in North America where they worked, that very few studies of effects of SO_2 were made for two decades. Even now, there is a strong emphasis in the United States on the observation of visible symptoms, and some reluctance to pay attention to invisible or physiological injury. This is in spite of the fact that much research since 1970 has pointed strongly to the existence of growth reductions in SO_2-polluted air. These reductions would not be easily recognised in the field in the absence of controls for comparison.

Effects of SO_2 on plant growth

Before 1970 very few experiments were performed using concentrations low enough to allow physiological injury, or growth reductions without visible leaf damage to be detected. The possibility of such effects had been raised by some experiments conducted by Bleasdale[4] in Manchester at about the time of the publication of the two reviews by Katz[1] and Thomas[3]. He grew ryegrass (*Lolium perenne* cv. S23) in two small greenhouses, one of which was ventilated with the ambient air at the site, and the other received air filtered to remove pollutants. The average daily concentration of SO_2 reached 0.10 — 0.30 ppm on only 5 in 192 days and never exceeded the value of 0.30 ppm which Katz had declared was the threshold for injury. Yet in eight out of nine experiments, the dry mass of the plants was greater in the unfiltered air, in three cases by over 50 per cent. Apart from a brief report published in *Nature* in 1952[4], Bleasdale's work received little publicity until the 1970s when many experimental fumigations of S23 ryegrass with SO_2 were conducted in different laboratories in the U.K.[5] The results of some of these are shown in Figure 1. Although there is considerable variation, the regression line is statistically significant ($P < 0.02$) and, taking into account the fact that the data were obtained by different people in different environmental conditions, Fig. 1 gives some encouragement to the view that a useful dose-response relationship for effects of SO_2 on growth can be obtained. The stimulation of growth by lower doses is a reality, for it has now been reported many times. It is not surprising that a supply of SO_2 can be beneficial. Sulphur is an essential element in plants and is not always in sufficient supply in the soil. However, while low doses

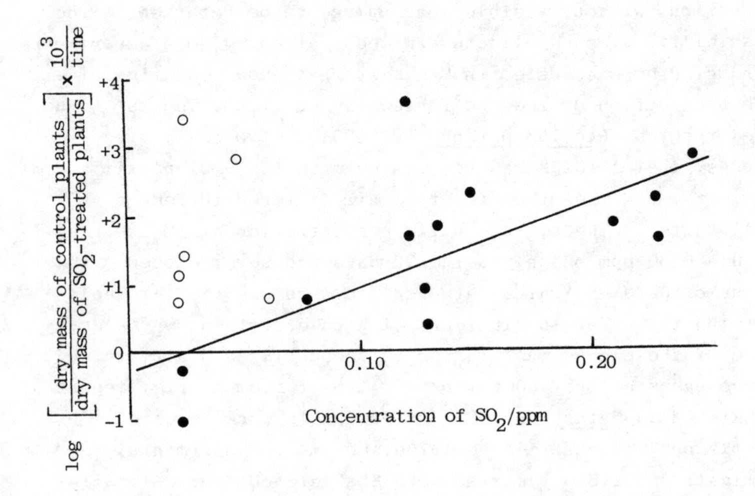

Figure 1. Closed circles represent effects on the shoot dry weight of ryegrass exposed experimentally to SO_2 pollution. Growth effects (ordinate) normalised for time in the manner adopted by Bell[6]. The open circles (not used in the fitting of the regression line) are the results of exposures to urban air of known SO_2 content. Regression significant at $P < 0.02$. Based on data from different experiments in two laboratories[7].

of SO_2 can correct sulphur deficiency, the leaf is not the normal route for the acquisition of a major part of the plant's sulphur, and the ions resulting from the entry of SO_2 into solution around leaf cells are potentially damaging.

Causes of variability in responses to SO_2

(1) Properties of fumigation chambers. Most of the published dose-response studies have been performed in plant growth chambers specially equipped for controlling and monitoring SO_2. The design of these chambers has, however, been very variable and often there is cause for concern about ventilation rate and velocity of air movement across the leaf surface.

Dose of SO_2 has usually been defined simply as the concentration in the atmosphere around the plants, or sometimes the concentration at the inlet. There has been a failure to recognise the importance of depletion within the chamber due to the uptake of SO_2 into the plants. The danger of defining dose in terms of inlet concentration is illustrated by the following data for soyabean (*Glycine max* L.) in a recent paper by Koziol[8]:

- (i) Hourly uptake of SO_2 into the leaves could be as high as 3 mg m^{-2}.
- (ii) The exposure chambers had a volume of 0.48 m^3, and thus during a treatment with 0.10 ppm SO_2 they would contain 0.137 mg of pollutant (at STP).
- (iii) The 10 plants contained in the chamber had a final leaf area of about 0.6 m^2, and their potential uptake of SO_2 was therefore 1.8 mg h^{-1}.

Koziol used these figures to illustrate the importance of a correct definition of exposure concentration. In some of his experiments, if he had quoted only the supply concentration (monitored at the chamber inlet, where the flow was 40 dm^3 min^{-1}), the actual exposure concentration would have been overestimated by a factor of 10. It is a cause of major concern that so much of the information in the literature on 'threshold concentrations' for injury by pollutants is subject to doubts about how these concentrations were defined.

Even if fumigation chambers are sufficiently well ventilated to maintain a stated concentration there is another important matter which can influence the dose of pollutant actually reaching the plant. This is the velocity of air movement past the leaves. If the velocity is too low, there is a high resistance to diffusion across the boundary layer, and although pollutant molecules are present in the surrounding atmosphere their rate of arrival at sensitive sites in the leaves may be reduced. Ashenden & Mansfield[9] exposed ryegrass to 0.11 ppm SO_2 in wind tunnels and found that the magnitude of response depended on wind speed. When the air flow across the plants was 10 m min^{-1} no adverse effects of this relatively high concentration of SO_2 could be detected. When, however, the wind speed was increased to 25 m min^{-1} the same concentration of SO_2 caused an appreciable inhibition of growth. The increase in wind velocity reduced the boundary layer resistance from 797 to 89 s m^{-1}. The higher of these resistances could be considered to act like a protective screen around the plant, reducing the diffusion of SO_2 molecules to sensitive sites in the tissues. Out-of-doors, boundary layer resistances would normally be lower than 100 s m^{-1}, but the air flow characteristics of many fumigation chambers are such that, typically, threshold tolerances of SO_2 have been determined with much higher resistances than this.

(2) Presence of other pollutants. The results of some experiments in which the growth of ryegrass was compared in filtered and unfiltered urban air are shown in Fig. 1, alongside the results of experimental fumigations with SO_2 on its own. It is clear that the toxicity of urban air is greater than would be expected from its SO_2 content. There are several other pollutants which may contribute to this toxicity, but evidence is growing that the most important of

them may be NO_2.

SO_2 and NO_x can be regarded as the most important primary gaseous pollutants in urban areas. A recent assessment by Fowler and Cape[10] for N.W. Europe has shown that they also occur together in areas generally regarded as 'rural', in concentrations which may affect plants. The relative proportions of NO and NO_2 in NO_x vary according to location. Here I shall discuss only the association of NO_2 with SO_2, since NO (because of its lower solubility) may be less important.

A paper by Tingey and colleagues[11] in 1971 first drew attention to the possibility that mixtures of SO_2 and NO_2 were much more toxic than would be predicted from their individual effects. They examined only the clearly visible, acute injury after 4-hour exposures to SO_2 and NO_2 individually and in various combinations. In these short treatments, 0.50 ppm SO_2 or 2.00 ppm NO_2 were the threshold concentrations for injury to tobacco (*Nicotiana tabacum* L.). However, when the individual concentrations exceeded 0.10 ppm in mixtures of the two, there was clearly visible injury which differed in detail from that caused by the gases on their own.

This was the first study to suggest that dose-response relationships obtained for individual gases under controlled conditions in the laboratory might be of little value in predicting effects in the field. Many other investigations have been made during the last 10 years, and in a recent review, Ormrod[12] was able to list 54 effects of pollutant mixtures which have now been defined, all but three of which involve SO_2. Synergism of the kind reported by Tingey *et al.*[11] has not been found in all cases, for sometimes effects are simply additive, and there are even instances of antagonism (particularly when SO_2 and O_3 occur together). However, enough evidence has now emerged to suggest that episodes of SO_2 + NO_2 could be particularly damaging in the field, and that the present information in the literature on individual effects of the two gases is of little value in predicting their action on crops, amenity plants and ecosystems.

Research conducted recently by some of my colleagues has pointed to a possible mechanism behind SO_2/NO_2 synergism at the cellular level[13]. Plants usually acquire nitrogen through their roots as NO_3^- or NH_4^+, but when NO_2 enters the leaves and dissolves in the extracellular water (which is held by surface tension in the walls of all plant cells) there will be a supply of both NO_3^- and NO_2^-. NO_2^- is known to be toxic even though it is an intermediate in the normal metabolic reduction of NO_3^-. The pathway and enzyme locations are thought to be

	enzyme	probable location
NO_3^-		
↓	nitrate reductase	cytoplasm
NO_2^-		
↓	nitrite reductase	chloroplasts
NH_4^+		

The rate control within this pathway is thought to be achieved by nitrate reductase, thus providing regulation of the amount of NO_2^- within the cell. When NO_2^- enters a cell following NO_x uptake into the leaves, the normal regulatory step is bypassed and toxic ions may accumulate. Nitrite reductase activity can, however, be stimulated by NO_x pollution and this must be considered as an essential first step towards detoxification. Wellburn et al.[13] found that the ability of leaves to form additional nitrite reductase in response to NO_2 pollution was strongly inhibited when SO_2 was also in the atmosphere (Table 1).

Table 1

Nitrite reductase activity (nmol nitrite reduced min^{-1} mg^{-1} chlorophyll) in extracts from leaves of ryegrass (*Lolium perenne* L.) grown in clean or polluted air for 20 weeks.

Variety of ryegrass	Control	SO_2	NO_2	$SO_2 + NO_2$
S23	135	113	452***	44***
S24	185	218	422***	171
S23 Bell	144	121	633***	52**
Helmshore	395	420	521*	186***

S23 and S24 are commercial varieties. S23 Bell and Helmshore[14] are clones selected for SO_2 resistance.

*P <0.05; ** P <0.01; ***P <0.001 (significance of difference from controls)
From Wellburn *et al.*[13]

This suggests that the enhanced damage when SO_2 and NO_2 are in combination is the result of the leaf's inability to prevent NO_2^- accumulating in the tissues as NO_2 enters from the atmosphere.

Because SO_2 is almost always accompanied by NO_2 in polluted air, prediction of their effects requires a series of dose-response studies with different concentrations in factorial combination. Very few researchers have sufficient facilities to pursue studies of this complexity at the present time, and one cannot foresee rapid progress towards the evaluation of effects in the field.

(3) Environmental conditions affecting growth. Studies of the effects of SO_2 on plants in the U.K. have often been performed in fumigation chambers subjected to natural illumination and ambient temperatures. In some of these experiments the growth responses to the pollutant seemed to be affected by weather conditions. For example, Bell *et al.*[15] found that the growth of ryegrass was inhibited (68% growth reduction) by 0.015 ppm SO_2 applied for 173 days under winter conditions, but such sensitivity has never been observed in summer by these or by other workers. Davies[16] made a study of effects of SO_2 on another grass (Timothy: *Phleum pratense* L.) and found that when growth was rapid, in high light intensities and long days, 0.12 ppm SO_2 had no influence. However, the same concentration applied in low light intensities and short days (i.e. a simulation of winter conditions) reduced growth by 50 per cent. These differences in the response of *P. pratense* have now been investigated in more detail[17] and there is strong evidence that slow growth enhances sensitivity to SO_2. When growth rate was purposefully slowed by reducing light intensity or temperature, pollution sensitivity seemed to be similarly increased (Figure 2). The results of these experiments make it abundantly clear that there is no critical concentration above which injury can be said to occur. Different doses of SO_2 can be tolerated under different climatic conditions. When rapid growth is favoured as in summer, *P. pratense* may not be affected by episodes of high SO_2 pollution (0.12 ppm as a daily mean is now found infrequently in the U.K., particularly in summer). On the other hand, effects which occur under less favourable conditions could be of importance in determining productivity, particularly if they involve reductions in survival ability under extreme climatic conditions. Recent studies at Newcastle University[18] have shown that plants exposed to SO_2 become more sensitive to frost injury. Figure 3 shows the substantial reduction in survival of polluted plants when exposed to sub-zero

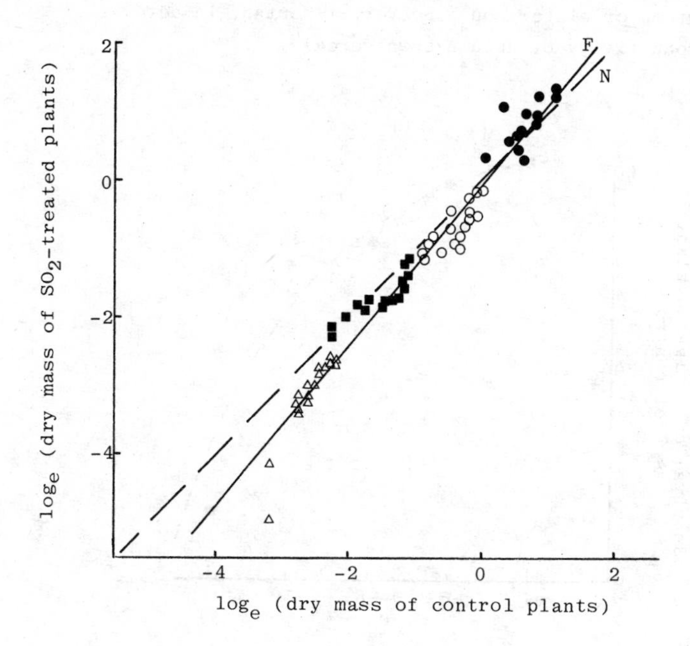

Figure 2. Relationship between growth rate and effect of SO_2 on dry mass increase of the grass Phleum pratense (Timothy). N is the "no effect" line and F is the line fitted to the data. Plants were fumigated for 44 days with 0.12 ppm SO_2 and the following environmental conditions were used to produce the different growth rates: ●, light of 400 $\mu E\ m^{-2}\ s^{-1}$, 19°C (night) and 30°C (day); ○, light of 400 $\mu E\ m^{-2}\ s^{-1}$, 12°C (night) and 26°C (day); ■, light of 100 $\mu E\ m^{-2}\ s^{-1}$, 19°C (night) and 30°C (day); △, light of 100 $\mu E\ m^{-2}\ s^{-1}$, 12°C (night) and 26°C (day). From Jones and Mansfield[17], and used with the permission of Applied Science Publishers Ltd.

temperatures. Frost susceptibility caused by exposure to SO_2 in autumn or winter could be very important in determining productivity of autumn sown cereals.

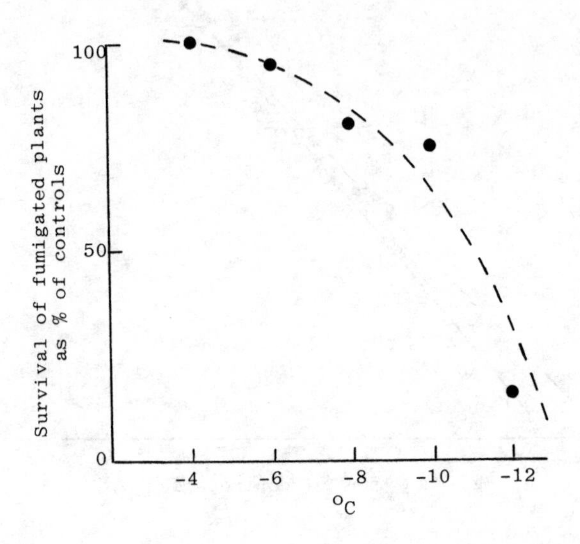

Figure 3. Survival of ryegrass (L. perenne cv. S23) after exposure to sub-zero temperatures. Controls were subjected to less than 15 µg SO_2 m^{-2}, and the fumigated plants were exposed for 3 weeks to 250 µg SO_2 m^{-2} (0.087 ppm). From Davison and Bailey[18] and reproduced with permission from Macmillan Journals Ltd.

(4) Plant nutrition. Sulphur is an important nutrient in plants because it is a constituent of amino acids and many metabolites, and because disulphide bonds and sulphydryl groups play vital parts in the structure and functioning of enzymes. The normal supply of sulphur for the plant is sulphate absorbed from the soil. In some agricultural areas the soil is deficient in sulphur and it is possible that SO_2 (and also H_2S) in the atmosphere could be able to remedy this deficiency. It is also possible that even when availability of sulphur from the soil is not limiting to plants, atmospheric sources could supply the leaves directly with their nutritional requirement, reducing the need for uptake from the soil.

Cowling and Koziol[19] have provided an excellent review of the many experiments performed over the last 50 years to ascertain whether atmospheric SO_2 can provide useful nutrition. There is much evidence that when soil supplies of sulphate are limiting to growth, SO_2 in the atmosphere can be beneficial. This appears to be particularly true when plants are growing rapidly under favourable conditions. The fact that SO_2 can be beneficial to plants needs, however, to be presented with caution. In the U.K., SO_2 pollution is greater in winter than in summer, i.e. at the time when plant growth is slowest. To benefit from SO_2 leaves must be able to use the supply of sulphur as it enters the cells. If metabolism is proceeding only slowly, toxic ions such as SO_3^{2-} and HSO_3^- will accumulate and damage will result. It is thus possible to put a mistaken emphasis on the beneficial effects of atmospheric SO_2, for even in situations where it may be of nutritive value for much of the time, short periods when metabolism cannot keep pace with its uptake will be followed by cellular injury, and subsequent loss of productivity.

Cowling and Koziol[19] have calculated the probable annual input of SO_2 into crops in rural areas. When the annual mean SO_2 concentration is about 0.01 ppm, the annual input could be as high as 38 kg ha^{-1}. This exceeds the known sulphur requirements of some crops.

(5) Evolution of resistance to SO_2. Populations of plants contain individuals with measurable differences in response to many environmental factors. The basis of much plant breeding is the initial selection of individuals with desired characters, with further selection from their offspring. This selection sequence can also be seen to occur without human intervention where a new environmental stress is imposed upon a plant population. The response of a population to the arrival of a toxic pollutant can be the appearance of tolerant individuals with surprising rapidity, e.g. 5-10 years. Some people express surprise that there is sufficient genetic variation in reactions to unnatural agents for evolution to occur in this way. There is, however, reason to believe that resistance to gaseous pollutants may be related to the ability to withstand other stresses, for example drought[20,21].

The evolution of resistance must sometimes be taken into account in the overall assessment of effects of a pollutant on growth and productivity. This is the case, for example, with the grasslands which occupy over 70% of mainland Great Britain and provide the food for large numbers of farm animals. Most other agricultural crops are grown with control over the spacing between individuals, but this is not the case with grasses except when a field is resown. As a newly sown field progresses to an established grass sward, there is severe competition between individuals of the same species, or between the different species that made up the seed mixture. Those plants which are more vigorous survive and squeeze out others, and a process of selection based on fitness in the particular environment (below and above ground) proceeds rapidly.

Estimates of growth reductions as a result of fumigations of a population of seedlings derived from commercially available grass seeds do not take into account these evolutionary changes. Predictions of the economic effects of pollution on grasslands therefore involve quite different considerations from those used with other crops, for example cereals (which are also grasses). The situation is further complicated by the knowledge that there are often, from a metabolic standpoint, measurable 'costs' of resistance[22]. For example, plants which are resistant to SO_2 in the atmosphere may have lower stomatal apertures than sensitive individuals. Partially closed stomata reduce the rate of uptake of SO_2, but they also restrict the supply of CO_2 for photosynthesis. This directly affects growth rate, and so there is a 'cost' of SO_2 resistance which can be readily understood.

Predicting effects on productivity: the future

Referring to the use of modelling, W.W. Heck[23] has recently written: "This term elicits a variety of responses from plant scientists. Some refuse to discuss the concept, pointing to extravagant claims of earlier modelling enthusiasts and their lack of success. Others think that modelling is the answer to all societal problems. The truth lies somewhere between these extremes but no serious scientific student of air pollution can afford to ignore the value or utility of modelling approaches".

Heck suggests that the best approach towards initial understanding of the responses of plants to air pollutants will be a detailed study of a particular species or plant type. This would enable critical responses (e.g. sensitive stages of growth) to be identified, and then these same effects could be sought in other plants. There is good sense behind these suggestions, but we must recognise that the time scale of any realistic approach of this kind is a long one. We must also recognise the dangers of drawing conclusions from experiments which are too small in their conception. Based on my own knowledge of the factors which control a plant's response to SO_2, namely the five outlined on the preceding pages, I could manipulate experimental conditions to produce quite different answers to a seemingly simple question, such as "does exposure to 0.06 ppm SO_2 inhibit growth?". I do not, therefore, find the confused state of the literature at all surprising. Many researchers in the past (and, it has to be admitted, some in the present), being unaware of the many factors controlling the responses to SO_2, have drawn conclusions which are unhelpful to those who turn to the literature for guidance when they have to make decisions. The statements of Katz[1] and Thomas[3] mentioned earlier must have had a considerable influence on those concerned with pollution control measures and legislation. It is now known that their estimated minimum SO_2 concentrations (circa 0.30 ppm) for causing injury may have been 10 times too high, at least for the crops we grow here in the U.K. and for our range of environmental conditions. The estimates may not, however, be incorrect for rapidly growing crops in some parts of the world. It is by performing experiments which are much broader in their conception than those of 30 years ago that we are now beginning to understand the variables involved.

<u>Other pollutants</u>

Much of what has been said above about SO_2 applies to other pollutants. Nitrogen is also an essential nutrient and beneficial as well as detrimental effects have also been found in experiments with NO_x pollution. O_3 does not, however, provide nutrition to plants and so its effects are not subject to this complicating factor. Errors in the experimental approach, particularly deficiencies in fumigation equipment, apply as much to other pollutants as to SO_2.

The literature must be studied critically, and it is strongly recommended that "toxicity thresholds" are not taken too seriously.

References

[1]M. Katz, Ind.Eng.Chem., 1949, 41, 2450.

[2]J. Stoklasa, "Urban und Schwarzenburg", Springer-Verlag, Berlin, 1923.

[3]M.D. Thomas, Ann.Rev.Pl.Physiol., 1951, 2, 293.

[4]J.K.A. Bleasdale, Nature, 1952, 169, 376.

[5]J.K.A. Bleasdale, Environ.Pollut., 1973, 5, 275.

[6]J.N.B. Bell, "Effects of Gaseous Air Pollution in Agriculture and Horticulture", ed. M.H. Unsworth and D.P. Ormrod, Butterworths, London, 1982, Chapter 11, p.225.

[7]T.A. Mansfield and P.H. Freer-Smith, Biol.Revs., 1981, 56, 343.

[8]M.J. Koziol, J.exp.Bot., 1980, 31, 1413.

[9]T.W. Ashenden and T.A. Mansfield, J.exp.Bot., 1977, 28, 729.

[10]D. Fowler and J.N. Cape, "Effects of Gaseous Air Pollution in Agriculture and Horticulture", ed. M.H. Unsworth and D.P. Ormrod, Butterworths, London, 1982, Chapter 1, p.3.

[11]D.T. Tingey, R.A. Reinert, J.A. Dunning and W.W. Heck, Phytopathology, 1971, 61, 1506.

[12]D.P. Ormrod, "Effects of Gaseous Air Pollution in Agriculture and Horticulture", ed. M.H. Unsworth and D.P. Ormrod, Butterworths, London, 1982, Chapter 15, p.307.

[13]A.R. Wellburn, C. Higginson, D. Robinson and C. Walmsley, New Phytol., 1981, 88, 223.

[14]J.N.B. Bell and C.H. Mudd, "Effects of Air Pollutants on Plants", ed. T.A. Mansfield, Cambridge University Press, 1976, Chapter 7, p.87.

[15]J.N.B. Bell, A.J. Rutter and J. Relton, New Phytol., 1979, 83, 627.

[16] T. Davies, Nature, 1980, 284, 483.

[17] T. Jones and T.A. Mansfield, Environ.Pollut.(Ser.A), 1982, 27, 57.

[18] A.W. Davison and I.F. Bailey, Nature, 1982, 297, 400.

[19] D.W. Cowling and M.J. Koziol, "Effects of Gaseous Air Pollution in Agriculture and Horticulture", ed. M.H. Unsworth and D.P. Ormrod, Butterworths, London, 1982, Chapter 17, p.349.

[20] W.E. Winner and H.A. Mooney, Oecologia, 1980, 44, 290.

[21] W.E. Winner and H.A. Mooney, Oecologia, 1980, 44, 296.

[22] M.L. Roose, A.D. Bradshaw and T.M. Roberts, "Effects of Gaseous Air Pollutants in Agriculture and Horticulture", ed. M.H. Unsworth and D.P. Ormrod, Butterworths, London, 1982, Chapter 18, p.379.

[23] W.W. Heck, "Effects of Gaseous Air Pollutants in Agriculture and Horticulture", ed. M.H. Unsworth and D.P. Ormrod, Butterworths, London, 1982, Chapter 19, p.411.

14
Epidemics of Non-infectious Disease

By P. J. Lawther
M.R.C. TOXICOLOGY UNIT, CLINICAL SECTION, ST. BARTHOLOMEW'S HOSPITAL MEDICAL COLLEGE, CHARTERHOUSE SQUARE, LONDON ECIM 6BQ, U.K.

ABSTRACT

Epidemics of non-infectious disease are often caused by exposure to industrial products, intermediates or by-products, either in the workplace or as a result of the contamination of a wider environment. Although the prime objective of research must be the recognition of the hazard and the evaluation of its magnitude so that illness may be prevented, close collaboration of clinicians, epidemiologists and toxicologists should lead to the acquisition of much knowledge of the mechanisms by which disease is caused. Catastrophes, though always regrettable, must be seen as experiments demanding careful analysis and exploitation. Many examples of different types of problem will be selected from the numerous epidemics from the time of the Schneeberg and Joachimsthal miners to the recent concern with contamination of the environs of Seveso by dioxin.

INTRODUCTION

The modest pretension of this necessarily brief contribution is that it might, by selecting for discussion a few of many epidemics of non-infectious disease, emphasize the need for collaboration of clinicians, epidemiologists, and toxicologists, in the difficult and vital tasks of identifying, assessing, forecasting and preventing long-term hazards to man from exposure to man-made chemicals in the environment. Hazards are recognized ultimately by clinicians and epidemiologists but, ideally, the toxicologist, who, by scrutiny of existing data and by further experiment, seeks to display the mechanisms by which substances exert their toxic effects, hopes as a result of his studies to be able to predict hazards and thereby avoid dangerous contamination of the environment. Clinical medicine, epidemiology and toxicology are disciplines which, practised in isolation, have obvious limitations; the need for collaboration in the study of environmental problems and the value of the results of such synergism are not so widely appreciated.

The clinical scientist is handicapped by the need, for ethical reasons, to limit his experiments on human volunteers to the study of mild, transient, easily reversible effects of exposure to suspect chemicals; moreover, his choice of subjects must be limited to healthy adults while populations at risk from contamination of the environment include the young, the old and the sick. The toxicologist who studies the effects of pollutants on animals suffers constraints which are not always recognized as being severe enough to diminish the relevance of his work to humans; the difference in reactions of various species of noxious stimuli is well recognized (though often ignored by those who seek to apply the results of the work of toxicologists) but the limits imposed by differences in life span and anatomy are less commonly recognized; the large increases in dose or concentration above those seen or expected in communal exposures, needed to produce effects in the animals' life time, may diminish the relevance of the findings; failure to recognize the importance of anatomical differences, especially of the respiratory tract in experiments on the effects of inhaled pollutants, have rendered invalid the application to man of many excellent experiments on animals. These observations or reminders are platitudinous in the context of this symposium, but when they go unheeded by those who set criteria and standards the economic consequences may be enormous.

Throughout the course of history there have been epidemics of disease due to contamination of the environment in which men work or where populations dwell; pollution may be from natural sources such as the hydrogen sulphide in Rotorua in New Zealand; man-made pollution may be accidental and catastrophic or it may be deliberate in ignorance of the harm it does or as a result of *laissez-faire*. The consequences of the contamination may be viewed merely as a matter for regret and a warning to reform; an additional, more positive, attitude is to regard such episodes as experiments that no one would be allowed to do and that offer unique opportunities for collaborative study by clinicians, epidemiologists and toxicologists. Retrospective scrutiny of past episodes must yield valuable lessons; perusal of Donald Hunter's classic work on *The diseases of occupation* and of the invaluable collections of papers in three issues of the *British Medical Bulletin* entitled 'Mechanisms of toxicity' (1969), 'Epidemiology of non-communicable disease' (1971), and 'Chemicals in food and environment' (1975), would convince anyone of the richness of this field of study. Industrial processes may cause disease by polluting the work-place and its environs; one is tempted to speculate that early epidemics may have been of silicosis in neolithic flint knappers and of farmers' lung among his agricultural contemporaries. Because of the clinical similarity of these two diseases and the commoner pulmonary tuberculosis and chronic bronchitis and emphysema, there must have been a delay in the course of time in the identification of the specific nature of these two diseases of occupation. The contaminated work-place could be seen as a most valuable laboratory in which experiments on humans have unwittingly taken place. The escape of prime products, such as lead, from industrial processes may contaminate the local surroundings and cause disease; similarly, the dissemination, accidental or deliberate, of by-products or waste material, such as mercury or cadmium, has caused infamous outbreaks of disease. Some waste products have been discharged in the genuine belief that they would cause no harm; later it has been found that harm has been caused by impurities in the waste products. The ash which collects in the heat exchangers in marine turbines powered by residual fuel oil seemed innocuous enough until it was found that workmen who were employed to remove it developed pneumonitis and other ills from the vanadium pentoxide which is a natural constituent of the pitch-like fuel used. There was no reason to believe at one time that the tailings from crocidolite workings could cause mesothelioma in the general population living near spoil banks. Poisoning of scientists in laboratories has produced invaluable evidence of hitherto unsuspected hazards; probably the smallest, yet not least relevant, epidemic was that which was reported by Edwards (1865, 1866) of the illness and deaths of two laboratory technicians who were working with dimethyl mercury (Frankland & Duppa 1863) at St Bartholomew's Hospital. The toxicity of mercury and its inorganic salts had long been known but the hazard posed by its organic compounds was not then recognized. The tragic story of the late recognition of the dangers of ionizing radiation by the injuries and deaths among the pioneers of radiology is well known and salutary. The side effects of therapeutic agents have sometimes been disastrous but may be regarded positively to have been experiments from which much can be learned. Almost incredibly, after the reports of the intense toxicity of dimethyl mercury mentioned above, diethyl mercury was used in 1887 for the treatment of syphilis with results which could have been forecast from Edwards' observations. In much more recent times the inclusion, for no good reason, of calomel in cathartic and antihelminthic preparations and in teething powders given to fractious infants led to epidemics of acrodynia or pink disease which would seem to have been an idiosyncratic response to the administration of a mercury salt. The consequences of the prolonged administration of arsenic in Fowler's solution for many chronic neurological disorders produced effects which many physicians have seen. Benzene has been used for the treatment of leukaemia. The sad effects of the ionizing radiation from the diagnostic use of thorotrast and the use of radiotheraphy in ankylosing spondylitis are well known and documented. The effects on the foetus of maternal treatment with thalidamide are well known. The follow-up of the victims of Hiroshima and Nagasaki has been a cardinal example of the beneficial examination of two calamitous episodes. Natural hazards can be used for the study of the potential toxicity of environmental contaminants. The effects of natural radiation have been investigated and the results applied to the effects of exposure to man-made radiation. Lead occurs in nature and the levels of lead in the blood of populations wholly unexposed to air-borne lead derived from petrol engines or from the products of industry have been used to compare with those of urban man. Mention has been made of Rotorua; in this small town in the North island of New Zealand the atmosphere is contaminated by hydrogen sulphide from geothermal sources, and deaths occasionally occur from acute poisoning in enclosed poorly ventilated spaces where the gas accumulates (lavatories attached to petrol stations are dangerous) but there are no signs of chronic ill effects of concentrations of the gas which many environmentalists would deem to be horrifying. It may be noted with regret that there is a reluctance to publish such negative findings which might provide welcome reassurance in days when almost every factor in the environment is said to be fraught with danger.

From these general comments it should be apparent that there is a wealth of material, the study of which will enable us to understand more fully real and suspected long-term hazards due to the presence of man-made chemicals in the environment. The examples which follow are selected from a vast field and no pretence is made that within the strict limits of this informal communication it is possible to do more than illustrate the general statements made above.

CARBON TETRACHLORIDE

One might hope to be allowed, in an informal presentation to a meeting for discussion, to speak from personal experience; some of the ideas presented here were conceived while recalling working in industry in wartime under appalling conditions in badly designed plant in which the need for high production took precedence over any consideration of industrial hygiene. A stage in the manufacture of a chemical vital to defence was the chlorination of benzanilide in solution in carbon tetrachloride. Spillages and leaks were frequent and the inadequate ventilation of the plant (the inadequacy was due to the need to observe the black-out regulations) allowed the accumulation of amounts of carbon tetrachloride much in excess of the 'maximum allowable concentration'; as a result of this most of the workers suffered from 'chronic carbon tetrachloride poisoning', a feature of which was severe nausea accompanied by lassitude and other demoralizing symptoms. Turnover of labour was high and efficiency low. The problem was serious enough to merit special investigation. Much was already known of the acute toxicity of this common and useful solvent; impairment of renal and hepatic function were known to be prominent features. Stewart & Witts (1944) reported the findings of the team which investigated the problem and found no evidence of impairment of renal or hepatic function among the workers but they did see signs of hypermotility and excess irritability of the gut and concluded that carbon tetrachloride in the concentrations breathed in the works probably caused the symptoms and signs by acting on the central nervous system. Other processes, including the manufacture of magenta and dinitrophenol, were carried out in other parts of the factory; process workers could be identified by their colours: magenta and bright yellow. All of those who worked on the process involving the chlorination of benzanilide had chloracne which was usually confined to the malar area of the face, though in some cases it was much more widespread.

POLYCYCLIC AROMATIC HYDROCARBONS

It could be said that the study of occupational cancer started when Percivall Pott wrote in 1775 of cancer of the scrotum in chimney sweeps and suggested that soot might be a cause of the tumours; Butlin (1892) reported his observations that exposure to mineral oil, pitch and tar was associated with similar tumours. Not surprisingly there was doubt about the identity of the carcinogen; the presence of arsenic, known to be carcinogenic, in coal tar was noted and blamed for the cancers by Pye-Smith (1913) and later by

Bayat & Slosse (1919) who concluded, 'Le cancer arsenical et le cancer du goudron sont identiques'. In 1915 Yamagiwa & Itchikawa produced cancer on rabbits' ears by painting them with extracts of coal tar, and thereafter Kennaway (1924, 1925) and Kennaway & Hieger (1930) isolated and identified the first known carcinogenic hydrocarbons, 1,2,5,6-dibenzanthracene and 3,4-benzpyrene, from pitch. There followed many reports that left no reasonable doubt that many skin cancers in industry were due to contamination of the body with tars and oils containing these or similar hydrocarbons. When causes were sought for the alarming increase in lung cancer among the general population, attention was drawn to the excess of the disease in towns; the air of most towns in western Europe was polluted by coal smoke, and Waller (1952) demonstrated the presence of 3,4-benzpyrene in town air. Since then the overwhelming cause of carcinoma of the bronchus has been shown to be the smoking of tobacco, especially in the form of cigarettes, but there remains an urban excess which some believe to be due to the presence of polycyclic hydrocarbons in town air. An appraisal of the role of these compounds in the aetiology of lung cancer was made possible by a study of the mortality of workers in the coal gas industry (Doll *et al*. 1965) which was supplemented by a survey of contamination of the air of retort houses by polycyclic aromatic hydrocarbons (Lawther *et al*. 1965). An excess of lung cancer (less than twofold) was seen among gas workers but this was far from being proportionate to the vast excess of 3,4-benzpyrene found in retort houses when compared with that determined in town air in the days when pollution by coal smoke was high. It would seem that too much importance has been ascribed to this class of compounds as causative factors of lung cancer in the general population.

The exhaust products of the diesel engine have been blamed for the rise in deaths from carcinoma of the bronchus. Kotin *et al*. (1955) demonstrated the presence of polycyclic hydrocarbons in soot from a maladjusted diesel engine; this finding was not surprising since these compounds may be found wherever carbon-containing fuels are burned inadequately. Subsequent experiments in which animals were exposed to exhaust from diesel engines were a failure because the animals died early from carbon monoxide poisoning. There was some irony in this failure since a valuable feature of the well adjusted diesel engine is that unlike the petrol engine it produces virtually no carbon monoxide. Again there seemed to be much to be gained by going to industry; surveys of pollution in London

Transport diesel bus garages were made (Commins et al. 1957) from which it was seen that the contribution of polycyclic hydrocarbons by diesel buses was small in comparison with that made by the coal fire and this was in accord with the findings of Raffle (1957) that there was no excess of lung cancer among workers in these garages. These figures are under annual scrutiny lest there be an effect of exhaust products which has not been manifested hitherto. The study of these carcinogenic compounds shows the value of collaboration between clinician, epidemiologist, chemist and experimental pathologist.

ASBESTOS

Asbestos is the name given to a group of fibrous silicates which are of great commercial value. Chrysotile, a fibrous form of serpentine which is a hydrated magnesium silicate, is also known as white asbestos and composes more than 80% of the World's output. Among the other varieties are the various types of amphibole silicates; these include crocidolite (blue), amosite (brown) and anthophyllite (white); there are many more. Although asbestos minerals have been known and used since ancient times, their use on a grand industrial scale dates from the discovery in Quebec and in Russia of large deposits of chrysotile about 100 years ago. There would seem to have been no good reasons to suspect that such chemically inert minerals would be harmful to man; the first case of what we now call asbestosis was observed in 1900 and later described by Murray (1907) but it was not until the late 1920s that there was enough evidence to establish an unequivocal relation between work with asbestos and the development of diffuse pulmonary fibrosis. This grim discovery prompted the enquiry which led to the classic report of Mereweather & Price (1930). Thereafter, many cases of asbestosis were reported from many parts of the industrial world. The Asbestos Industry Regulations followed in 1931 and dust control was enforced in 'scheduled' processes. Inhalation of any of the asbestos minerals can give rise to phenomena which are of little or no clincial significance; the presence of asbestos bodies in the sputum signifies nothing more than that asbestos fibres have been inhaled; likewise the presence of fibro-fatty pleural plaques commonly seen in asbestos workers is thought to be of no clinical import; calcified pleural plaques, which are endemic in parts of Finland where the soil contains much anthophyllite, are often seen in workers who have inhaled asbestos and are held by many clinicians to be harmful only when they are widespread enough to cause restriction of movement of the lungs. In the mid-1930s came the suggestion that carcinoma of the bronchus was more common in asbestos workers and this suggestion was investigated, confirmed and quantified by Doll (1955) and has been the subject of much later work. The excess incidence of carcinoma of the bronchus in asbestos workers has been shown to be greatest in those exposed over long periods, extending back to the time when concentrations were high (Newhouse 1973), and more recently the importance of cigarette smoking as a truly synergistic factor has led some to ask whether asbestos is carcinogenic in its own right or whether it merely enhances the carcinogenicity of tobacco smoke. During the war there was a report from Germany claiming that pleural 'cancers' were more frequent in asbestos workers, but it was much later (1956) that Wagner and his colleagues noted the occurrence of large numbers of cases of pleural and peritoneal mesotheliomata associated with exposure, often slight, to crocidolite in the northwest region of Cape Province (Wagner 1960). These malignant tumours have since been shown to occur often very many years after exposure to minute amounts of crocidolite and are not dependent on the presence or absence of asbestosis. They are rarely seen in chrysotile miners. Clearly, there seems to be some special property of crocidolite (an iron silicate) which enables it to cause these rare and fatal tumours. At first much attention was paid to the fact that iron was substituted for magnesium and that the difference in pathogenicity was 'chemical'; later differing electron densities in the various fibre crystals were blamed. More recently, differences in fibre size and other physical dimensions between crocidolite and the other types of asbestos have been studied: some animal experiments have shown that the tendency for chrysotile, when injected intrapleurally, to cause mesothelioma can be enhanced by altering the size and shape of the particles so that they come to resemble crocidolite fibres.

This interest in the physical rather than chemical nature of asbestos fibres in relation to their pathogenicity has been justified and furthered by some recent accounts of the geographical distribution of mesotheliomas: Baris (1975) reported an analysis of 120 'pleural mesotheliomas and asbestos pleurisies due to environmental asbestos exposure in Turkey'. Included in this series was a group of 39 patients from the villages of Karain and Urgup in which asbestos had not been found; however, the villagers use a local 'white soil' for numerous purposes and this geologically complex mixture contains, among many other minerals, volcanic glass fibres with dimensions similar to those of some crocidolite fibres. The geology, epidemiology and pathological findings in this remote region are receiving much attention. (It is of great interest to note that chest disease is familial in this region where the very name of the village Karain means 'pain in the chest'.) Das et al. (1976) have reported and discussed five cases of mesothelioma of the pleura occurring in rural India among people with no history of exposure to asbestos but who were engaged in the sugar cane industry; they discuss possible aetiological factors and it is obvious that this part of the world merits the same careful investigation as is going on in Anatolia. These matters, in which clinicians, epidemiologists, and pathologists must work together, are of much more than academic interest: when the dangers attendant on the use of asbestos, especially crocidolite, were fully realized, advice was given to seek and use other materials with similar physical properties but which were not known to be hazardous. The 'latent period' for the development of mesothelioma may be long and the causation may be related to physical rather than chemical properties of inhaled fibres; great caution is therefore needed before assurances of the safety of asbestos substitutes and other man-made fibres can be given. But it must be pointed out that the results of intrapleural injection of test animals with comparatively large doses of suspect fibres and powders must not be taken to mean that the effects seen are necessarily applicable to man. This whole topic may be used as a model of those problems the solution of which needs the application of many skills and much wisdom.

MERCURY

Mercury is an ancient metal and there have been many accounts of epidemics of poisoning by the metal and its inorganic salts (all occupational) since Pliny described the disease of slaves who worked in mercury mines. The symptoms of poisoning include insomnia, shyness, nervousness, dizziness and tremor. The 'non-specific' - or, more properly, common - nature of the symptoms lead one to the certain belief that many cases are missed and diagnosed erroneously as psychoneuroses. The volatility of metallic mercury is popularly underestimated and one has wondered if some odd behaviour of some laboratory workers may be due to inhalation of mercury vapour derived from spilled metal which has been ignored. The diagnosis of mercury poisoning is made easier when there is a clear history of work with the metal and if the poisoning is severe enough to cause salivation and gingivitis and signs of the nephrotic syndrome. Even so, there have been patients in whom other diagnoses have been made (and wrong treatment given) through failure to take an adequate occupational history. There is therefore reason to believe that some epidemics of mercury poisoning have gone undetected. Nevertheless, perusal of accounts of poisoning among miners, gilders, mirror makers, surgeons, hatters, thermometer makers, meter menders, and detectives is rewarding if only to remind one of the importance of clinical observation in the practice of epidemiology. Pink disease, previously mentioned, has been eradicated since the use of poisonous teething powders has been discontinued. Unlike the symptoms of poisoning by mercury and its inorganic compounds, many of which disappear after exposure ceases, the damage caused by exposure to organic mercurials tends to be more serious and is less often reversible. The sad fate of the late laboratory technicians at St Bartholomew's Hospital has already been mentioned. The toxicology of the organomercurials is complex: an admirable summary is that by Magos (1975). Some organic compounds of mercury are excellent fungicides and are used for the treatment of many seeds. When the seeds are sown, the organomercurials are broken down and are made biologically inactive in the soil. Because free movement of mercury from the roots of the plant to the leaves or grain is prevented by some biochemical mechanism, there is no accumulation of mercury in the terrestrial food chain. But obviously poisoning can occur in the process of formulating the seed dressing and applying it and by the consumption of treated grain dressed for sowing. Several severe outbreaks of poisoning have occurred by all these routes. Mercury can enter the terrestrial food chain if birds or animals which have eaten dressed seeds are consumed. A classic early account of the symptoms and signs of poisoning by industrial exposure to methyl mercury was given by Hunter *et al.* (1940). They include generalized ataxia, dysarthria and gross constriction of the visual fields which can proceed to blindness. Memory and intelligence are said to remain relatively unimpaired in less severe cases. Fortunately, industrial cases of poisoning are now rare as the toxicity of the compounds is well recognized. But, tragically, there have been, since Hunter's description of his cases, many more opportunities to study the symptoms, signs, and pathology of the effects of alkyl mercury. Epidemics involving large numbers of people have been caused by eating bread made from wheat and other grains which had been treated with methyl or ethyl mercury. The biggest tragedy so far recorded (Bakir et al. 1973) occurred in Iraq during the winter of 1971/72 when, as a result of eating bread made from dressed grain, 6000 patients were admitted to hospital and more than 500 died. There had been previous epidemics from the same cause in Iraq, Pakistan and Guatemala, and smaller episodes had been reported from other countries, but only in the 1971/72 epidemic in Iraq were there quantitative studies in which exposure was related to observed clinical effects (Bakir et al. 1973; Kazantzis et al. 1976a; Mufti et al. 1976; Shahristani et al. 1976). These workers severally were able to assess the dose-response relation for various clinical signs and symptoms.

In the aquatic environment, inorganic mercury compounds may be methylated by the action of certain bacteria and enter the food chain via fish. There have been several epidemics of poisoning by methyl mercury among fish-eating peoples but the two major epidemics occurred in Japan in Minimata Bay (Katsuna 1968) and in Niigata (Niigata Report 1967) and were caused by the industrial release of methyl and other mercury compounds into Minimata Bay and into the Agana River after which the mercury was absorbed by fish which were then eaten. By 1971 as many as 260 cases of poisoning by methyl mercury had been reported in Minimata and Niigata, and of these 55 were fatal. More than 700 cases of poisoning had been identified in Minimata by 1974; more than 500 had been identified in Niigata. Again, these epidemics were intensively studied and, as a result, unique assessments of dose-response relations could be calculated for methyl mercury (Swedish Expert Group 1971). In addition, it was seen that the foetus was more susceptible to methyl mercury than was the mother. Details of these findings and of observations made of other epidemics of poisoning by mercury compounds are summarized admirably in W.H.O. Environmental Health Criteria, volume 1 (1976), from which yet again one may learn the inestimable value of the exploitation of catastrophes by clinicians, epidemiologists and toxicologists working together.

DIOXIN

The compound 2,3,7,8-tetrachlorodibenzo-p-dioxin (TCDD) is extremely toxic and very stable. An enormous amount of work has been done on its toxicology in animals; in addition to producing chick oedema in chickens it has legion harmful effects. It has the teratogenic, foetotoxic and porphyrogenic effects which are well documented. A sensitive test of its presence is the production of hyperkeratotic lesions when painted on the ear of a rabbit. It and related chlorinated dibenzodioxins were synthesized by Tomita et al. (1959); it is not used commercially but it is found as a contaminant when 2,4,5-trichlorophenol is synthesized by hydrolysis of tetrachlorobenzene at high temperatures. The compound 2:4:5-trichlorophenol (2,4,5-TCP) is used to make 2,4,5-T and 2,4-D which are commonly used as effective herbicides. Dioxin may be disseminated in minute quantities as trace impurities in these compounds. There have, however, been several catastrophic releases of dioxin as a result of plant explosions when the exothermic reaction producing 2,4,5-TCP has got out of control. The most recent episode, which has set public health officials unprecedented problems and research workers a unique field for clinical and toxicological investigation, occurred at Seveso near Milan in 1976 (Giovanardi 1977) when an area of 3-4 km^2 was contaminated by a major leak from a factory making 2,4,5-TCP. The area was severely polluted by TCDD; part of the area has been evacuated. The legion effects of dioxin on man have been noted from careful observations following contamination. There was a famous episode when three horse arenas in Missouri were contaminated by spraying the ground with waste oil which had been contaminated (Carter et al. 1975; Kimbrough et al. 1977); 57 horses died. A girl 6 years old who had played in an arena became severely ill with many signs and symptoms. She was intensively studied until she recovered completely. Other persons were affected and were studied carefully (Beale et al. 1977). From 1960 to 1968 mixtures of 2,4,5-T which were contaminated with TCDD were sprayed as a defoliant over large areas of Vietnam; Cutting et al. (1970) instituted studies to see whether the inadvertent exposure of the general population had given rise to an excess of birth defects. Not surprisingly, this exercise was fraught with difficulties and the interpretation of the findings is still the subject of debate. An increase in liver tumours in Vietnam had been reported by Tung (1973). The lesions seen in the more acute cases exposed to the higher concentrations resulting from local plant failure have been varied and severe. An almost common feature has been the production of chloracne. This skin lesion can be produced by exposure to several chlorinated naphtholic compounds but dioxin seems to be particularly powerful as a cause of chloracne (May 1973). There are several cases of contamination of the environment with a relatively innocuous compound (2,4,5-TCP) in which a micro-contaminant has been the cause of widespread disease. I am reminded of my wartime experience of chloracne when manufacturing a chlorinated benzanilide derivative and the indication would seem to be a retrospective analysis of such compounds to see if their capacity to cause chloracne was due to contamination by TCDD or similar compounds. There will remain the fascinating problem for the toxicologist to solve: by what mechanism is a peripheral lesion such as chloracne produced by such a compound.

POSTSCRIPT

Indulgence is sought for the platitudinous presentation of mere selection of ill-assorted instances where the collaboration of clinician, epidemiologist, and toxicologist has been essential to the elucidation of important problems in preventive medicine. The protection of the environment must surely depend upon such liaison. But lest one be tempted to think that all problems of environmental contamination might be elucidated in the near future, one is reminded that the cause of the ancient epidemic of lung cancer among the miners of Schneeberg could not have been revealed by the most careful investigations by clinician, epidemiologist and toxicologist until Marie Curie had discovered radium. Our liaison must embrace all disciplines and we must seek and hope for some new discovery which will throw more light on the mechanisms of toxicity so that we shall be able to predict hazards and so avoid them.

REFERENCES

F. Bakir, S.F. Damluji, L. Amin-Zaki, M. Murtadha, A. Khalidi, N.Y. al-Rawi, S. Tikriti, H.I. Dhahir, T.W. Clarkson, J.C. Smith & R.A. Doherty, 1973, Science, N.Y. 181, 230-241.

Y.I. Baris, 1975, Hacettepe Bull. Med. Surg. 8, 165-185.

A. Bayet & A. Slosse, 1919, Bull. Acad. med. Belg. 29, 607.

M.G. Beale, W.T. Shearer, M.M. Karl and M.M. Robson, 1977, Lancet i, 748.

British Medical Bulletin 1969 Mechanisms of toxicity. Br. med. Bull. 25, 3.

British Medical Bulletin 1971 Expidemiology of non-communicable disease. Br. med. Bull. 27, 1.

British Medical Bulletin 1975 Chemicals in food and environment. Br. med. Bull. 31, 3.

H.T. Butlin, 1892 Br. med. J.i, 1341.

C.D. Carter, R.D. Kimbrough, J.A. Liddle, R.E. Cline, M.M. Zack Jr,

W.F. Barthel, R.E. Koehler & P.E. Phillips, 1975 Science, N.Y. 188, 738.

B.T. Commins, R.E. Waller & P.J. Lawther, 1957 Br. J. ind. Med. 14, 232-239.

R.T. Cutting, T.H. Phuoc, J.M. Ballow, M.W. Benenson & C.H. Evans, 1970 Congenital malformations hydatidiform moles and stillbirths in the Republic of Vietnam 1960-1969.

P.B. Das, A.G. Fletcher Jr. & S.G. Deodhare, 1976 Aust. N.Z. J. Surg. 46, 218.

R. Doll, 1955 Br. J. Ind. Med. 12, 81.

R. Doll, R.E. Fisher, E.J. Gammon, W. Gunn, G.O. Hughes, F.H. Tyrer & W. Wilson, 1965 Br. J. ind. Med. 22, 1-20.

G.N. Edwards, 1865 St. Bart's Hosp. Rep. 1, 141.

G.N. Edwards, 1866 St. Bart's Hosp. Rep. 2, 211.

E. Frankland & B.F. Duppa, 1863 J. chem. Soc. (N.S.) 1, 415.

A. Giovanardi, 1977 In Proceedings of the Expert Meeting on the Problems Raised by TCDD Pollution. Milan, 30 September and 1 October 1976, pp. 49-50.

D. Hunter, R.R. Bomford & D.S. Russel, 1940 Q. Jl Med. 9, 193.

D. Hunter. 1975 The diseases of occupation, 5th edn. London: English Universities Press.

M. Katsuna, (ed.) 1968 Minamata disease, Japan: Kumamoto University.

G. Kazantzis, A.W. al-Mufti, A. al-Jawad, Y. al-Shahwani, M.A. Majid, R.M. Mahmoud, M. Soufi, K. Tawfiq, M.A. Ibrahim & H. Debagh, 1976 In World Health Organization Conference on Intoxication due to Alkyl Mercury Treated Seed, Baghdad, 9-13 November 1974, p. 37. Geneva: World Health Organization. (Suppl. to Bull. Wld Hlth Org. 53).

G. Kazantzis, A.W. al-Mufti, J.F. Copplestone, M.A. Majid & R.M. Mahmoud, 1976 In World Health Organization Conference on Intoxication due to Alkyl Mercury Treated Seed, Baghdad, 9-13 November 1974, p. 49. Geneva: World Health Organization. (Suppl. to Bull. Wld Hlth Org. 53).

E.L. Kennaway, 1924 J. Path. Bact. 27, 233.

E.L. Kennaway, 1925 Br. J. med. J. ii, 1.

E.L. Kennaway & I. Hieger, 1930 Br. med. J. ii, 1.

R.D. Kimbrough, C.D. Carter, J.A. Liddle, R.E. Cline & P.E. Phillips, 1977 Archs envir. Hlth 32, 77-86.

P. Kotin, H.L. Falk & M. Thomas, 1955 A.M.A. Archs ind. Hlth 11, 113-120.

P.J. Lawther, B.T. Commins & R.E. Waller, 1965 Br. J. ind. Med. 22, 13-20.

L. Magos, 1975. Br. med. Bull. 31, 241-245.
G. May, 1973 Br. J. ind. Med. 30, 276-283.
E.R.A. Mereweather & C.W. Price, 1930 Report on effects of asbestos dust on the lungs and dust suppression in the asbestos industry, London: H.M.S.O.
A.W. al-Mufti, J.F. Copplestone, G. Kazantzis, R.M. Mahmoud, & M.A. Majid, 1976 In World Health Organization Conference on Intoxication due to Alkyl Mercury Treated Seed, Baghdad, 9-13 November 1974, p.23. Geneva: World Health Organization. (Suppl. to Bull. Wld Hlth Org. 53.)
M. Murray, 1907 Departmental Committee on Compensation for Industrial Diseases, Cmnd 3495, p. 14; Cmnd 3496, p. 127. London: H.M.S.O.
M.L. Newhouse, 1973 Ann. occup. Hyg. 16, 97-107.
Niigata Report 1967 Report on the cases of mercury poisoning in Niigata. Tokyo: Ministry of Health and Welfare.
P. Pott, 1775 Chirurgical observations relative to the cataract, the polypus of the nose, the cancer of the scrotum, the different kind of ruptures and the mortification of the toes and feet. London: Hawes, Clarke & Collins.
R.J. Pye-Smith, 1913 Proc. R. Soc. Med. (Clin.) 6, 229.
P.A.B. Raffle, 1957 Br. J. ind. Med. 14, 73-80.
H. al-Shahristani, K. Shihab & I.K. al-Haddad, 1976 In World Health Organization Conference on Intoxication due to Alkyl Mercury Treated Seed, Baghdad, 9-13 November 1974, p. 105. Geneva: World Health Organization. (Suppl. to Bull. Wld. Hlth Org. 53.)
A. Stewart & L.J. Witts, 1944 Br. J. ind. Med. 1, 11-19.
Swedish Expert Group 1971 Nord. hyg. Tidskr. Suppl. 4, p. 65.
N. Tomita, S. Ueda & N. Narisada, 1959. J. pharm. Soc. Jap. 79, 186-92.
T.T. Tung, 1973 Chirurgie 99, 427-436.
J.C. Wagner, 1960 In Proc. Pneumoconiosis Conf., Johannesburg. 9-24 February 1959 (ed. A.J. Orenstein), p. 373. London: Churchill.
R.E. Waller, 1952 Br. J. Cancer 6, 8-21.
K. Yamagiwa & K. Itchikawa, 1915 Mitt. med. Fak. K. jap. Univ. 15, 295.

ACKNOWLEDGEMENT

This article was first published in Proc. R. Soc. Lond. B. 205, 63-75 and is reproduced with kind permission of The Royal Society.

15
Systems Methods in the Evaluation of Environmental Pollution Problems

By P. Young
DEPARTMENT OF ENVIRONMENTAL SCIENCES, UNIVERSITY OF LANCASTER, LANCASTER LA1 4YQ, U.K.

Introduction

One of the first steps in the scientific method is the formulation of a hypothesis, usually in the form of a mathematical model of some kind. The confidence that the scientist has in this model will vary considerably, depending upon the nature of the application. In the case of natural environmental pollution problems, the degree of confidence is often low: whilst the scientist may be able to specify numerous mechanisms which could be operative in the system, he often finds it difficult to define which of these mechanisms will be dominant under the specific circumstances encountered in the problem under study.

This lack of confidence is compounded by the problems associated with the planning and execution of *in situ* experiments on the system. Planned laboratory experiments may help to clarify certain aspects of the system behaviour but they can rarely define unambiguously the exact nature of this behaviour in the natural environment. Faced with such a dilemma, the scientist must resort to passive observation or "monitoring" of the system during its normal operation. At the very best, such monitoring will include the acquisition of data from partially planned experiments, as when tracer exercises are used to evaluate the transportation and dispersion of a controlled release of tracer material into a river or estuarine system. At the worst, monitoring will be restricted by logistic and cost considerations, so that the resultant data set can only reveal limited aspects of the system behaviour.

Such difficulties are exacerbated if it is the *dynamic* behaviour of the system which is important to the resolution of the problem under consideration. The objective of the present paper is to introduce new methods of time-series and systems analysis that can help in the evaluation and solution of this kind of "badly defined" system problems.[1,2] Needless to say, the analysis can be no better than the data set on which it is based and it may well be that the information/data-base may be so meagre that little sensible analysis is possible. Nevertheless, the proposed approach can provide a system of "checks and balances" which can help the analyst and model builder at least to identify difficulties which are inherent in the limited data set. In this manner, it is possible to avoid problems such as "over-parameterisation", "surplus unvalidated content" and unjustified confidence in the resulting model.[1,2]

An Illustrative Example : Modelling the Dispersion of Pollution in a River System

Before discussing a general approach to the analysis and modelling of poorly defined systems, let us consider a simple yet practical example. Here the behaviour is relatively well defined, but difficulties of model formulation, identification and estimation can still arise if conventional modelling methods are used without taking note of certain results in systems theory.

Classical hydrodynamic analysis associated with dispersion in flowing media is most often related to the seminal work of G.I. Taylor[3] on flow in pipes and involves a mathematical representation in the form of a single dimensional, partial differential equation, usually known as the Fickian Diffusion equation.[3,4] Although many textbooks subscribe to the use of this equation in a river context, reasonable arguments can be presented which suggest that it is not necessarily an appropriate model for dispersion in natural channels and streams.

Recently, an alternative theory has been suggested[5] which assumes that the major dispersive mechanism resides not in classical dispersion coefficient D but in the "residence-time" T associated with the aggregative effect of all "dead-zone" phenomena in the river between any two sampling points. Such dead-zone behaviour can arise because of holes and irregularities in the bed and bank, meanders and pool-riffle sequences, all of which tend to characterise natural channels and lead to transient retention and mixing processes in the stream.

In chemical engineering terms, this aggregated dead-zone (ADZ) model is a combination of "plug flow", to account for the translational effect introduced by the main river flow mechanism, and a "continuous-stirred-tank-reactor" (CSTR) to describe the ADZ behaviour. The model is most easily derived by considering a mass balance over a single reach of the river. For the purpose of this analysis, it is assumed that the reach length has been chosen such that the dispersional behaviour can be described by a single such ADZ model. If the flow through the reach is denoted by Q m^3 sec^{-1} and the volume of the ADZ (normally less than the reach volume) is V m^3, then, under the assumption of complete mixing, the relationship between the input (upstream) concentration $c_i(t)$ and the output (downstream) concentration c(t) is obtained from mass conservation considerations in the form,

$$\begin{array}{ccccc} \text{rate of change of mass} & = & \text{mass flow in} & - & \text{mass flow out} \\ \frac{d[Vc(t)]}{dt} & = & Q\,c_i(t-\tau) & - & Q\,c(t) \end{array} \quad (1)$$

or, if Q and V are initially assumed constant,

$$\frac{dc(t)}{dt} = -\frac{1}{T}\,c(t) + \frac{1}{T}\,c_i(t-\tau), \quad (2)$$

where τ is the pure time delay introduced to allow for the translational effects of the river flow, and $T = V/Q$ sec is the ADZ "residence time".

A discrete-time equivalent of equation (2), which is important to our subsequent discussion, can be expressed in the following form,

$$c_k = a\,c_{k-1} + b\,c_{i,k-\delta} \quad (3)$$

where the subscript k denotes the value of the variable c or c_i at the kth sampling instant, while δ is the discrete-time equivalent of the pure time delay τ in (2). In other words, if the sampling interval is T_s sec, then $\tau = \delta.T_s$. Finally, if the tracer material or pollutant is assumed to be conservative then, with minor assumptions, a and b can be related to

T and T_s by the following expressions

$$a = \exp(-T_s/T) \; ; \; b = 1-a \tag{4}$$

Figure 1 compares the modelling results obtained from an ADZ analysis of radioactive tracer data from Copper Creek, Virginia,[6] with the results obtained by Fischer[7] using the Fickian Diffusion representation. Clearly, in this example, the ADZ model (with a two reach structure; i.e. two ADZ elements such as equation (3) in series) is better able to explain the data than the more conventional distributed parameter model.

If, on the basis of results such as those in Figure 1, we assume that the ADZ model is a more appropriate description of dispersion in a river system than the more conventional Fickian Diffusion equation†, then several questions need to be answered before we can use the model. Since the theory is of comparatively recent origin, some of these questions cannot be answered at this time. For example, it is not yet possible to associate measurement of the ADZ residence time directly with the physical characteristics of the reach; nor is it possible to predict any changes in T that may arise with changes in flow conditions. It is possible, however, to consider one important aspect of the model: namely, how do we choose the length of river reach appropriate to a single ADZ description; or, equivalently, given concentration measurements at two points in the river, how many ADZ elements are required to provide an adequate dynamic description of the data? These questions can be answered by reference to measurements obtained either from a tracer experiment conducted on the river or from the analysis of natural tracer data (e.g. salinity) obtained during normal monitoring. Here, we will consider the former, although the method of analysis can equally well be applied to the latter, since it does not require any specific form of input $c_{i,k}$.

---

† In general terms this assumption must itself be tested more thoroughly before proceeding with the analysis.

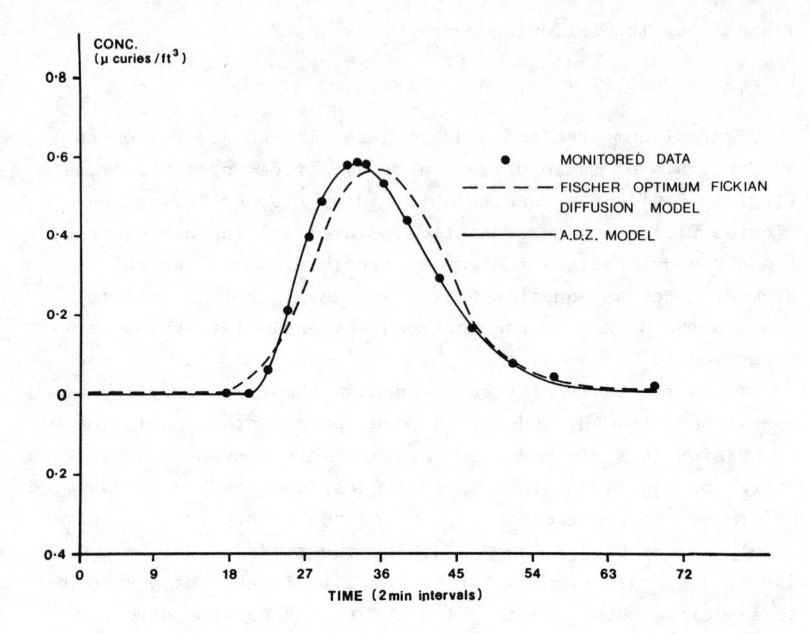

Figure 1 Models for Pollutant Dispersion in Copper Creek, Virginia

(A) Fickian Diffusion Equation:

$$\frac{\partial c(x,t)}{\partial t} + U \frac{\partial c(x,t)}{\partial x} = D \frac{\partial^2 c(x,t)}{\partial x^2}$$

with the mean velocity coefficient U and the longitudinal dispersion coefficient D estimated by the routing method of Fischer[7].

(B) Aggregated Dead Zone Model:

2 ADZ compartments in series with parameters estimated by recursive instrumental variable approach[5,8].

The results of a typical tracer experiment carried out on the River Wyre near Lancaster are shown in Figure 2. Here trace A is the "input" concentration of the fluorescent dye Rhodamine W.T., as measured at a point 0.5 km downstream of the pulse (or "gulp") injection point (which was situated close to the effluent outfall of the Scorton Treatment Plant). Trace B is the output concentration measured some 1.5 km further down the river. These data were obtained in a microcomputer controlled monitoring exercise associated with a field course for first year students taking the Environmental Sciences Degree Course at Lancaster. Figure 2 and subsequent figures in the Section were obtained directly as printer output from an APPLE microcomputer. All time-series analysis discussed in the paper was also carried out in the microcomputer.

In order to decide how many ADZ elements are appropriate to the model, we could look at the physical nature of the river and choose a reach length which seemed intuitively sensible. For example, bearing in mind the nature of the ADZ model, we might consider that a reach length of 0.5 km is reasonable and, therefore, select a model with three reaches. As we shall see, such a model might well be considered appropriate, in the sense that it could be fitted to the data with only a small residual error. But such a decision on reach length seems somewhat subjective and we might question whether the resultant model is necessarily the "optimum" model or whether some other model may not be more appropriate to the explanation of the experimental data.

Such considerations raise several theoretical questions which are important to dynamic systems theory but which we cannot consider rigorously in this paper. These questions relate to general concepts such as the "identifiability" of the model from the experimental data and whether the input signal (in this case the impulse or gulp of dye) is "sufficiently exciting" to allow for the identification and estimation of the selected model.[8,9] Here, however, we will restrict our discussion to the results of some exercises in time-series analysis which demonstrate the relevance of concepts such as these in relation to the data shown in Figure 2.

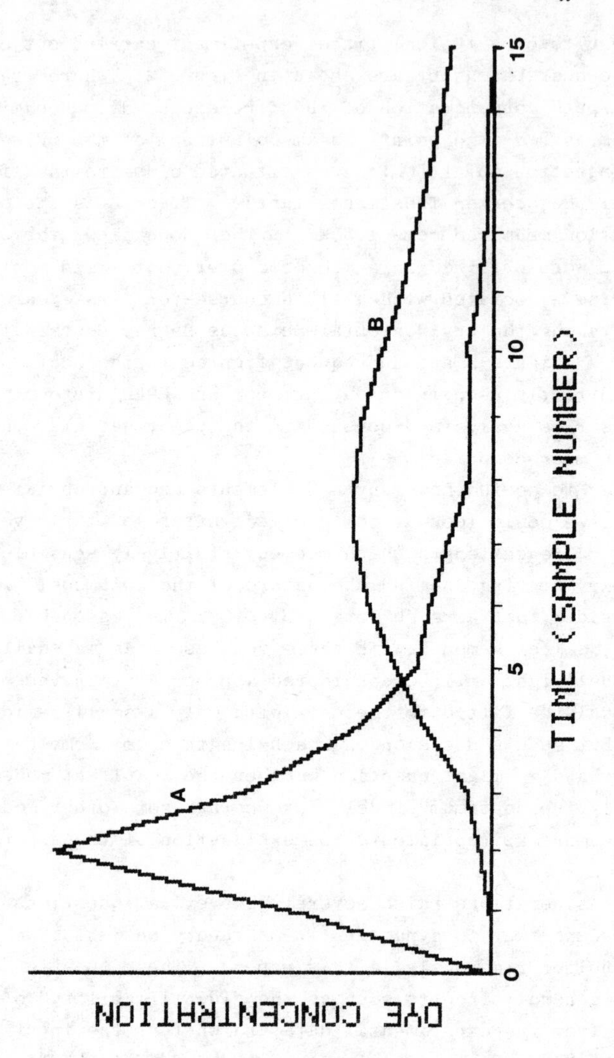

Figure 2. Dye Concentration Data obtained from an Experiment on the River Wyre, Lancashire (direct microcomputer output)

Young[1,2] and Kalman[10] have both stressed the importance of limiting the effects of the model builder's "intuition" or "prejudice" on model construction. And both have stressed the need to proceed from "data → model" in as objective a manner as possible. Of course, complete objectivity is difficult in practice; but at least the analyst can attempt to limit the strength of his prejudice in relation to model construction. This is made easier if he follows a systematic procedure which, as we have indicated in the introduction, provides a series of "checks-and-balances" at all stages of the modelling activity.

Bearing this philosophy in mind, it is necessary to reach some objective decision on what specific model of the general ADZ type is most appropriate to the data in Figure 2. Here we will use the approach suggested by Young et al.[11]: this involves repeated estimation of parameters in different order models and the computation of various statistical measures which help to define that model order which is best suited to the experimental data. In basic terms, the objective of the analysis is to find the simplest description of the data and, as such, it is in sympathy with the concept of simplicity discussed by Popper.[12,1,2]

The two statistics of primary concern to the model order identification procedure are the "coefficient of determination" (R_T^2), and the "error-variance-norm" (EVN). R_T^2 is defined quite simply as

$$R_T^2 = 1 - \sum_{i=1}^{N} e_i^{\,2} \Big/ \sum_{i=1}^{N} (c_i - \bar{c})^2$$

where e_i is the model residual (fitting error) and $\bar{c}$ is the mean value of c_i. Thus R_T^2 is a normalised measure of the degree of model fit; with unity value if there is a perfect fit, and zero value if the model is failing to explain the data at all. An exact definition of EVN is given elsewhere[11] but, for our present purposes, it can be considered as a normalised measure of the variance associated with the parameter estimates: a low relative value is indicative of well defined, low variance estimates; a much higher value indicates poor definition and high variance, both of which tend to characterise an over-parameterised model. As a result, when the model order is higher than can be justified by the data, the EVN tends to increase sharply, often by several orders of magnitude. For this reason, it is usual to quote the EVN in terms of its natural logarithm, ln. EVN. Such a large increase is normally indicative of model ambiguity: i.e. a number of different model parameter sets can equally well explain the data.

Table 1 shows the main results obtained from this kind of identification analysis for the tracer data in Figure 2; the microcomputer output for the second order model is shown in Figure 3. It is clear that the second order model is preferable from these results: the model fit is only marginally worse than that for the third order model and the EVN is much lower indicating much better defined parameter estimates. In other words, although the third order model fits the data very well, it appears over-parameterised so that the marginal increase in R_T^2 cannot be justified by the increase in model complexity. Thus the 3rd order model can be considered to have a degree of "surplus content" not relevant to the experimental data.

Table 1

MODEL ORDER	R_T^2	ln. EVN
1	0.949	-5.860
2	0.998	-6.423
3	0.999	-3.305

The nature of the parameterisation used in the model and shown in Figure 3 is not important to the present discussion. It will suffice to note that the model structures tested are general, linear, discrete-time dynamic models, which can be considered as multi-order extensions of the model (3), but which are not restricted to the multiple ADZ type.[1,8] However, within the levels of uncertainty associated with estimates (the

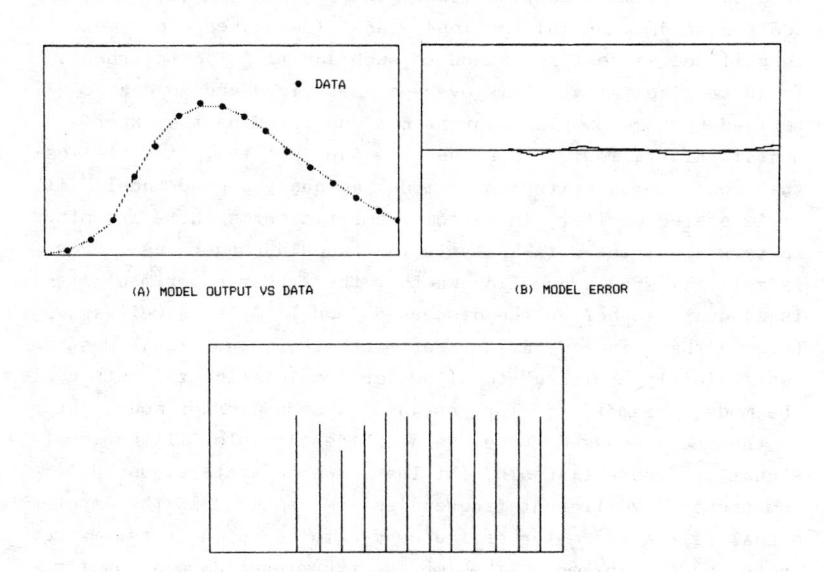

Figure 3. Estimation Results for a Second Order (2 reach, ADZ) Model of the Wyre Data (direct microcomputer output)

approximate standard errors are shown in parentheses next to the estimates), the second order estimated model is equivalent to two ADZ elements in series, each with a residence time of 3.73 sampling intervals, or about 30 minutes for the sampling interval of 8 mins. used here.

The lessons of this simple example are obvious. The limited data available in this experiment can only support the hypothesis of a second order, two reach model characterised by three parameters. Any more complex model would need to be justified by additional data or information. And if a higher order model is utilised without reference to such additional data, then it is quite likely that it is over-parameterised and so characterised by some surplus content not justified by the experimental information. This despite the fact that it satisfies that most common criterion of model adequacy a good model "fit".

As stated earlier, in system theoretic terms there are other factors about the data analysis that should concern us. For example, we should question whether the tracer experiment itself is adequate to define the dispersion model. It is well-known (e.g.[8]) that, in a dynamic experiment, the input signal must be "sufficiently exciting" to allow for complete identification of the model parameters: for instance a second order model, such as the one discussed above, is not "identifiable" if the input signal is a pure sinusoid; at least two separate sinusoidal components of different frequencies are required in the input signal if the parameter estimates are to be uniquely estimated. In the present example, the impulse type input does allow for identifiability in these terms† but it is certainly not an optimum input[9]. This raises the idea of introducing better experimental planning procedures in relation to these and similar dynamic experiments.[13]

Of course, such ideas of optimum experimental planning are only relevant in relation to those situations where experimental planning is possible. However, since most environmental modelling must be based largely on monitored data, we must also ask ourselves whether such passively obtained data contain sufficient information to allow for complete model identification and estimation. Obviously, the answer will depend upon the nature of the system, the model and the data. All that

---

† Although, strictly, it is not persistently exciting.[8]

can be said in general terms is that the analyst should attempt to evaluate his modelling results with these factors in mind. And, as a golden rule, he should be extremely wary of using large and complex models when his data-base is small, as is so often the case in pollution modelling. Otherwise, he may well be attaching too much weight to his prior prejudices and may have more confidence in his model predictions than is justified by his knowledge of the system.

A General Approach to Modelling of Pollution Processes in the Natural Environment

In relation to the general problem of pollution modelling, the implications of the simple example discussed in the previous section are quite far reaching. In specific terms, the ADZ model is simple and yet there is initial evidence that it may have potential for wider application in areas such as streamflow modelling, dispersion and flow through 'macropores' in groundwater, and the dispersion of air pollutants. In more general terms, if we allow for non-conservative pollutants (which is straightforward), then the multivariable equivalent of equation (1) is a state-space model of the form

$$\frac{d\underline{x}}{dt} = A\underline{x} + B\underline{u}$$

where $\underline{x} = [x_1 \; x_2 \; .... \; x_n]^T$ is an nth order vector of state variables (e.g. the concentrations of pollutants); $\underline{u} = [u_1 \; u_2 \; .... \; u_m]^T$ is an mth order vector of inputs which give rise to variations in the state variables (e.g. sources of pollutants); and A,B are n x n and n x m matrices of coefficients or parameters which characterise the interrelationships between the variables and define the dynamic behaviour of the model. Such models are commonly used to describe chemical, biological and ecological behaviour in the natural environment and have been used extensively in pollution modelling. Beck and Young[14], for example, have utilised this kind of model to characterise biochemical oxygen demand (BOD) - dissolved oxygen (DO) relationships in natural streams and there are many other examples in the literature of more complex pollution models.

The time-series analysis procedures discussed in the last section can also be generalised to handle these more complex multivariable situations, and this has led the author[1,2] to suggest a general procedure for modelling badly defined systems which has direct relevance to the modelling of pollution in the natural environment. The various steps in this procedure will not be discussed in detail in this paper since they have been discussed fully in the quoted references. Nevertheless, it is worth noting that the procedure makes extensive use of "recursive" methods of time-series analysis in which model parameter estimates are updated sequentially whilst working serially through the time-series data. An example of recursive estimation is given in Figure 3, where the recursive estimate of one of the second order model parameters is shown in (C). The value of such estimation is that it allows the analyst to relax the normal assumption that the model parameters are time-invariant and, in this manner, he can investigate the possibility of non-stationarity and non-linearity in the model.[1,2] This is particularly useful in nonlinear model structure identification.

Whilst the recursive methods of time-series analysis are sophisticated in estimation terms, they are not excessively complicated and are quite easy to use in practice. This is emphasised by the fact that the analytical results discussed in the last section were all obtained from an APPLE Microcomputer programmed in the BASIC language. As a result, it is possible to use the programs for "on-line" data processing in the field, with the recursive estimation allowing for the updating of model parameters as the data are received from the measuring instruments. Indeed, the microcomputer has been used in the dispersion example as a step in the development of a robust, on-line data acquisition and processing facility for use in field applications.

A diagram of the overall modelling procedure is shown in Figure 4, where it is seen that, following the careful definition of the objectives of the modelling exercise, the analysis proceeds systematically through various stages, with allowance for feedback and iteration if satisfactory results are not achieved at first. This is in sympathy with a simple

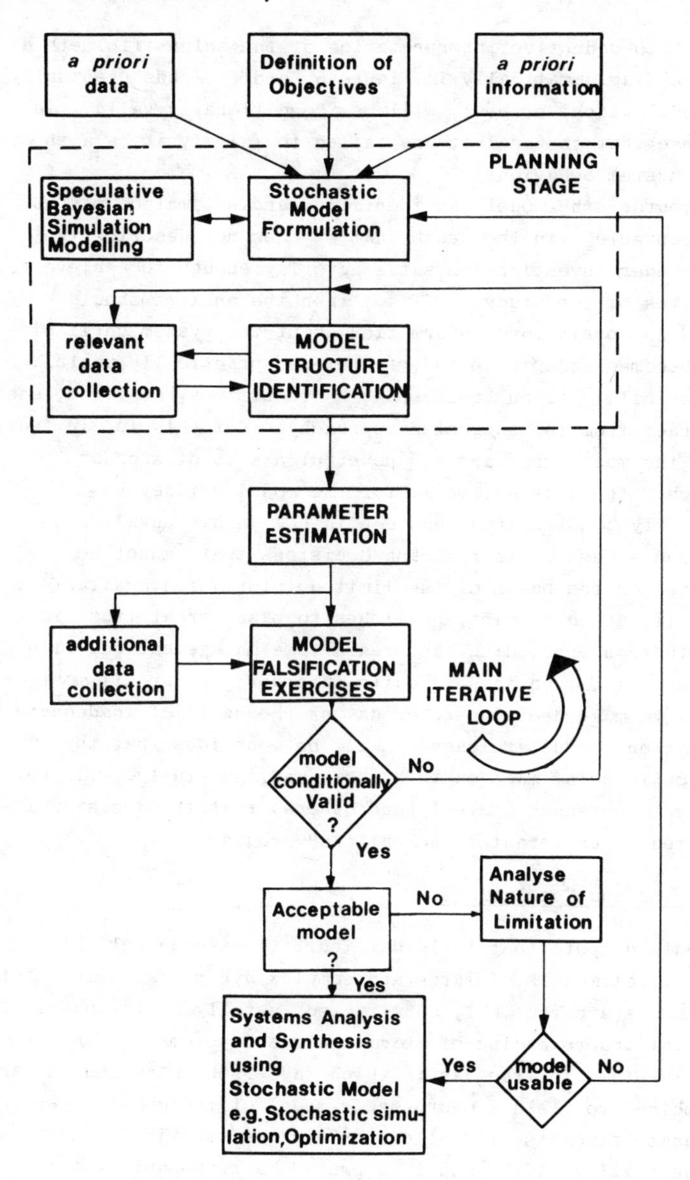

Figure 4. Simplified Block Diagram of the Approach to Modelling Badly Defined Dynamic Systems Proposed in the Paper

hypothetico-deductive interpretation of the scientific method as shown diagrammatically in Figure 5, and it leads eventually to a model which, at best, will be "conditionally valid", in the sense that the analyst has failed to falsify it as a theory of the system behaviour.[12]

Of course the model, as identified and estimated, may not be "acceptable", in the sense that it does not describe the system under investigation satisfactorily enough to achieve the objectives of the study. If so, then the analyst should proceed to obtain more information about the system until his model becomes acceptable and, therefore, practically useful. The possibility of an "unacceptable" result may, at first, seem to detract from the suggested approach. But this is, in fact, one of the most important and powerful assets of such an approach: the alternative is for the model builder unconsciously to place too much confidence in his unvalidated model and so use it in reaching decisions that cannot be justified on the basis of the limited information available to him. If, at this point, he wishes to place great trust in his intuition (prejudice?) and reach some management decision on this basis, then this is quite in order; after all most of us have to make decisions everyday on the basis of inadequate information. But at least he will be conscious that the decision is being made on this basis and, hopefully, will not attempt to persuade himself (and others) that the decision is justified on unwarranted "scientific" grounds

Discussion and Conclusions

Few would dispute that it is difficult to develop reliable mathematical models of large and complex dynamic systems. Yet such models are routinely constructed, usually on the basis of a partial understanding of micro-scale subsystems and an intuitive feel for the way in which these subsystem models can be combined to yield a macro-scale model of the complete system. Inadequate attempts at "validation" or "calibration" then follow, and the ability of the model to provide a reasonable fit to limited sets of *in situ* data is often considered as a primary justification for assuming holistic validity.

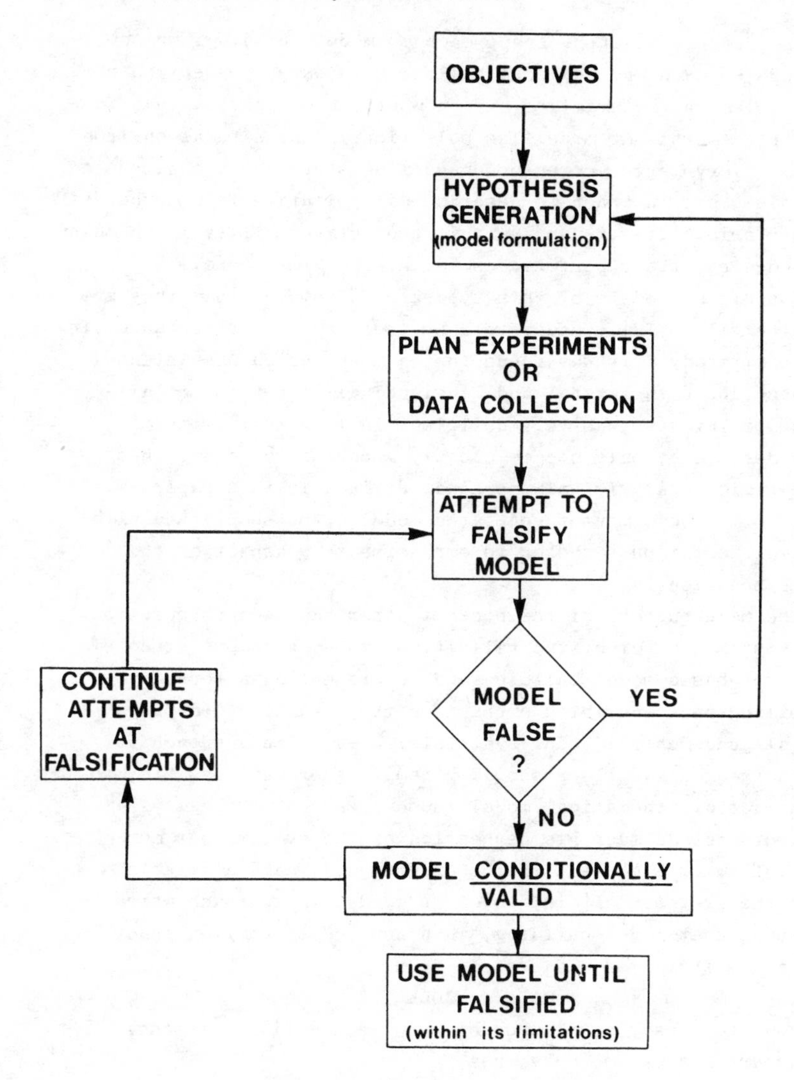

Figure 5. The Model Building Approach Considered within a Popperian Interpretation of the Scientific Method

The dangers of such an approach to model building, which depends so much on the model builder's untested perception of the system characteristics, are nowhere more apparent than in current attempts at modelling pollution in the natural environment. Very large, computer-based models have been constructed on this basis in order to describe air pollution in a number of major World cities. And yet, if predictive ability is anything to judge by, their performance is, at best, mediocre.

Too often, models of this type are so complex that they are not themselves fully understood by the user. For instance, in a recent study[15] it was found that a model, which was intended to describe both lateral and longitudinal dispersion of airborne pollutants, was very deficient in longitudinal terms. Such deficiency only became clearly apparent, however, when time-series analysis, of the kind outlined in this paper, was applied to the computer generated model outputs obtained with special test inputs chosen to emphasise this aspect of the model behaviour.

The main purpose of the present paper has been to question the wisdom and scientific validity of these attempts at complex, computer-based model building and to present an alternative modelling approach which emphasises the need to avoid prejudicial judgement in model synthesis. It is an approach in which the model builder should attempt, wherever possible, to construct a mathematical model whose size and complexity reflects not so much his perception of the system, but more the information content in the experimental data available from the system. It is, in other words, an approach which stresses that model building, when applied to complex, poorly defined systems, should proceed from

$$\text{DATA} \rightarrow \text{MODEL}$$

without too heavy a reliance on prior conceptions of model structure and parameter values.

Of course, pollution models constructed in this manner may well not describe the system adequately for all purposes. But this merely emphasises the poverty of the data collected in many studies of environmental pollution and our lack of understanding of the macro-scale behavioural patterns. The main laboratory for the scientist concerned with the study of environmental pollution is the environment itself. Analytical and research laboratories in universities, research institutions and industry can clearly help in understanding the fundamental processes involved in environmental pollution. But they cannot hope to define these processes completely. It is one thing, for example, to analyse a complicated chemical engineering process or a sophisticated aerospace vehicle that has been constructed by man and whose basic structure is well known; it is quite another to model ill-defined natural environmental systems where the luxury of the planned in situ experiment is the exception rather than the rule.

Finally, the fact that the simple example in this paper has illustrated the inherent difficulties of analysing data, even from planned experiments, should serve to emphasise the enormous problems which face the scientist who attempts to model environmental pollution on the basis of monitored data. It is hoped that the caveats presented here will be heeded by research workers who are confronted with this difficult but challenging task. The advice may not help them to solve their problems but it may prevent them drawing unjustified conclusions on the basis of inadequate information.

References

1. P.C. Young, in "Modeling, Identification and Control in Environmental Systems", G.C. Vansteenkiste (ed.). North Holland, Amsterdam, 1978, p.103.
2. P.C. Young, in "Uncertainty and Forecasting of Water Quality", M.B. Beck and G. van Straten (eds.). IIASA/ Pergamon, Oxford, in press.
3. G.I. Taylor, Proc. Roy. Soc., 1954, A223, 446.
4. H.B. Fischer, Int. Jnl. Air and Water Pollution, 1966, 10, 443.
5. P.C. Young and T. Beer, Centre for Resource and Environmental Studies, Australian National University, Rep. No. AS/R42, 1980 (sub. for publication in modified form).
6. R.G. Godfrey and B.J. Frederick, U.S. Geological Survey, Paper 433-K, 1970.
7. H.B. Fischer, Proc. A.S.C.E., Jnl. Sanit. Eng. Div., 1968, 94, 927.

8. P.C. Young, "An Introduction to Recursive Estimation", Springer-Verlag, Berlin, 1982, in press.
9. G.C. Goodwin and R.L. Payne, "Dynamic System Identification: Experiment Design and Data Analysis", Acad. Press, New York, 1977.
10. R.E. Kalman, Int. Jnl. Policy Anal. and Inf. Syst., 1979, 4, 3.
11. P.C. Young, A.J. Jakeman and R. McMurtrie, Automatica, 1980, 16, 281.
12. K.R. Popper, "The Logic of Scientific Discovery", Hutchinson, London, 1959.
13. A.J. Jakeman and P.C. Young, Proc. 4th Biennial Conf., Simulation Soc. of Australia, 1980, 248.
14. M.B. Beck and P.C. Young, Proc. A.S.C.E., Jnl. Env. Eng. Div., 1976, 102, EES, 909.
15. L.P. Steele, "Recursive Estimation in the Identification of Air Pollution Models", Ph.D. Thesis, Australian National University, 1981.

16
Organometallic Compounds in the Environment

By P. J. Craig
SCHOOL OF CHEMISTRY, LEICESTER POLYTECHNIC, PO BOX 143, LEICESTER LE1 9BH, U.K.

I Introduction

(1) Scope

Organometallic compounds usually enter the environment following use as commodities, although there are instances of the formation of these compounds in the environment (e.g. mercury methylation). Some organometallic products are applied directly to the environment as biocides, in anti-fouling paints or in petrol. Others reach the wider environment indirectly (e.g. leaching from organotin-based PVC stabilizers). In general it is necessary to consider not only the direct toxicity of the compound but the toxicities of possible metabolites at points other than those of initial application. Formation of organometallic compounds in the environment is important because the organic derivatives are of greater toxicity than their parent inorganic metals or ions. Hence the complete cycling or transport of the original organometallic compound in the environment must be considered.

The scope of this chapter includes only compounds with metal-carbon σ bonds. Metal organic complexes and inorganic reactions are covered elsewhere. Some metalloidal elements which form methyl derivatives in the environment are included (e.g. arsenic and some aspects of selenium chemistry). Legal aspects and organometallic compounds in the workplace are also covered elsewhere.

Certain vetinary uses of organometallic compounds do have environmental impact and this will be noted. Theoretical aspects of environmental methylation (biomethylation) will be mentioned only where directly relevant to the environmental situation. A recent discussion of this topic is available[1].

With regard to environmental methylation an important difficulty has concerned the low levels at which some of these compounds are formed. Methyl mercury is generated at the part per billion level (ng g^{-1}), and the methyl tin derivatives have been observed at the part per trillion level (pg g^{-1}). These observations are significant; continuous formation of organometallic compounds in the environment even at low levels may result in food chain effects leading to much higher concentrations in organisms. The main analytical problem seems to be the use of techniques that work well on standard test materials for operation in a mixed environmental matrix. Analytical problems are noted as they arise.

(2) Source Material

Tin, lead and mercury, whose organometallic compounds are of most environmental concern, have been reviewed in detail in monographs. Environmental concentrations, commercial uses, toxicity, persistence and methylation of the metal and compounds are discussed.[2-4] The cycling of metals and organometals in the environment has been reviewed recently,[5] and is also covered in single volume work.[6] Biological methylation is discussed in a number of series.[7] Organometallic compounds in the environment was the subject of an American Chemical Society (A.C.S.) monograph.[8] The U.S. National Academy of Science and the A.C.S. have published information on lead and mercury in the environment.[9,10] The World Health Organization has produced a series of monographs on Environmental Health Criteria, and mercury, lead and tin organometallic compounds have been covered.[11-13] Organometallic compounds used for medicinal or vetinary use are noted in the respective Pharmacopoeiae.[14-16] Other organizations have reported on mercury in the environment.[17-18] There is a very recent monograph on lead in the environment.[19] The degradation properties of organotin compounds have also been reviewed very recently.[20,21]

(3) Units and Abbreviations

S.I. units are used where concentrations are quoted. These are related to "trivial" units, as follows.

S.I. Unit	Trivial Unit
$\mu g\ g^{-1}$ ($mg\ kg^{-1}$)	ppm
$ng\ g^{-1}$ ($\mu g\ kg^{-1}$)	ppb
$pg\ g^{-1}$	ppt
$\mu g\ dm^{-3}$	$\mu g\ L^{-1}$
$\mu g\ m^{-3}$	

Concentrations unless otherwise noted refer to "dry weight" of the matrix (e.g. sediments or fish).

Where an element or a simple compound is being referred to the full terminology is used (e.g. lead or methyl mercury). Complex materials are abbreviated (e.g. Me_3Sn^+, Me_4Pb). Methyl mercury refers to monomethyl mercury ($MeHg^+$) and dimethyl mercury appears in full. MeCoB12 is methyl cobalamin; PVC is polyvinylchloride. Alkyl groups are abbreviated conventionally. Analytical techniques are listed in the normal abbreviated form (e.g. atomic absorption is written AA.)

II Organotin Compounds

(1) Use and Toxicity

The chief uses of organotin compounds are as stabilizers for polyvinyl chloride (PVC) and as biocides and both lead to the introduction of those compounds to the environment.
As a result their direct toxicity and their breakdown pattern are of importance in any consideration of their environmental role. Dialkyl tin compounds are used for polymer stabilization, and triorganotin species have biocidal properties. Production of organotin compounds is more than 30,000 tonnes annually. A number of reviews on the organotin compounds exist.[20-24]

Table 1

Some Uses of Triorganotin Compounds

Compound	Use
Ph_3SnOAc	Fungicide, anti-fouling paint.
Ph_3SnOH	Fungicide, anti-fouling paint
Ph_3SnCl	Anti-fouling paint
Ph_3SnF	Anti-fouling paint
$Ph_3SnSCSNMe_2$	Anti-fouling paint
$Ph_3SnOCOCH_2Cl$	Anti-fouling paint
$Ph_3SnOCOC_5H_4N_3$	Anti-fouling paint
$Ph_3SnOCOCH_2CBr_2COOSnPh_3$	Anti-fouling paint
Bu_3SnOAc	Anit-fouling paint
$Bu_3SnOCOPh$	Disinfectant
Bu_3SnCl	Rodent repellant
Bu_3SnF	Anti-fouling paints
$(Bu_3Sn)_2O$	Fungicide; bacteriostatic (wood or stone preservative)
Bu_3Sn adipate	Anti-fouling paint
Bu_3Sn methacrylate (copolymer)	Anti-fouling paint
$Bu_3SnOCOCH_2CBr_2COOSnBu_3$	Anti-fouling paint
Bu_3Sn (naphthenate)	Bacteriostatic (wood preservative)
$(Bu_3Sn)_3PO_4$	Bacteriostatic (wood preservative)
$(cC_6H_{11})_3SnOH$	Insecticide (orchards)
$(cC_6H_{11})_3Sn$—N(N=C, C=N) (triazolyl ring)	Insecticide
$[(PhC(Me)_2CH_2)_3Sn]_2O$	Insecticide, acaricide
$Et_3Sn(pOC_6H_4Br)$	Nematocide
$(CH_2{=}CH)_3SnCl$	Herbicide
Me_6Sn_2	Insecticide

Table 2

Some Uses of Diorganotin Compounds

Compound	Use
$[nOct_2Sn(C_4H_2O_4)]_n$ (maleate polymer)	PVC stabilisation (including food contact)
$nOct_2Sn(SCH_2COOiOct)_2$	PVC stabilisation (including food contact)
$Bu_2Sn(OAC)_2$	Polyurethane catalysts; cold curing of silicones
$Bu_2Sn(OCOiOct)_2$	Polyurethane catalyst; cold curing of silicones
$Bu_2Sn(OCOC_{11}H_{23})_2$ (dilaurate)	PVC stabilisation; de-worming of chickens, catalysis
$Bu_2Sn(SCH_2COOiOct)_2$	PVC stabilisation
$[nBu_2Sn(C_4H_2O_4)]_n$	PVC stabilisation
$Bu_2Sn(OCOCH{=}CHCOOOct)_2$	PVC stabilisation
$Bu_2Sn(SC_{12}H_{25})_2$	PVC stabilisation
$Bu_2Sn(OCOC_{12}H_{25})_2$	Catalysis
$[Bu_2SnO]_n$	Catalysis
Bu_2SnCl_2	Glass strengthening; precursor for SnO_2
Me_2SnCl_2	Glass strengthening; precursor for SnO_2
$Me_2Sn(SCH_2COOiOct)_2$	Heat stabiliser and food contact (PVC)
$(BuOCOCH_2CH_2)_2Sn(SCH_2COOiOct)_2$	PVC stabilisation

Table 3

Some uses of Monoorganotin Compounds

Compound	Use
$MeSn(SCH_2COOiOct)_3$	PVC stabilisation
$MeSnCl_3$	Glass strengthening, SnO_2 precursor
$Bu(SCH_2COOiOct)_3$	PVC stabilisation
$BuSnCl_3$	Glass strengthening
$(BuSnS_{1.5})_4$	PVC stabilisation
$BuSn(OH)_2Cl$	Catalysts for transesterification
$(BuSn(O)OH)_n$	Catalysis
$OctSn(SCH_2COOiOct)_3$	PVC stabilisation
$BuOCOCH_2CH_2Sn(SCH_2COOiOct)_3$	PVC stabilisation

(i) Triorganotin Compounds as Biocides

The more important feature is the nature of the organic group, rather than the anionic group. The main biocidal applications are summarized in Table 1, and account for about 8000 tons of organotin production.

It can be seen from Table I that triorganotin compounds are introduced direct into the environment. Hence knowledge of their toxicity and degradation properties in the environment has been an important research goal in recent years. Toxic effects for organotin compounds are at a maximum when three organic groups are present (i.e. R_3SnX). Usually maximum toxicity occurs for R = methyl to butyl; higher trialkyl tin species are of little toxicity. The basic cause of acute triorganotin toxicity arises from disruption of various mitrochondrial functions through membrane damage, disruption of ion transport and inhibition of ATP synthesis. Choice of alkyl group in the triorganotin biocides represents a balance between the excessive phytotoxicity of the lower alkyls and insufficient toxicity for the higher alkyl compounds.

Decay in the environment is most likely to occur through absorption of U.V. light, biological action or chemical means. Essentially, decay occurs by stepwise loss of the organic group and the final product has been postulated as the non-toxic tin dioxide (cassiterite). This decay to progressively less toxic materials is an attractive environmental feature of organotin products. It does appear however, that the picture is not quite as simple as this; for example, many studies suggest that by no means all of the introduced tin is finally converted to the oxide. *In vitro* studies suggest that decay does occur to the oxide, but when biological or light induced decay studies are carried out in a realistic environmental model decay is often incomplete and apparently stable organotin products may remain. This tends to be confirmed by the recent detections of methyl and butyl tin species at low concentrations (ng dm^{-3})in environmentally dispersed situations, e.g. rivers, rainwater etc.. This suggests a greater stability in the environment than some laboratory studies on pure compounds have implied. That pure triphenyl-, dibutyl- and dialkyl-tin compounds undergo stepwise decay under UV light to inorganic tin under abiotic conditions should be assessed from this point of view.[20,21]

Biological metabolism of trialkyltins has been modelled by the rat liver microsomal monooxygenase system and occurs mainly by hydroxylation of carbon atoms on the organotin.[25-27] The hydroxy metabolites decay to dialkyltin compounds by tin-carbon bond cleavage with further carbon hydroxylation and cleavage to tin (IV) oxide. In the environment similar biological oxidations are believed to decay other trialkyltins to inorganic tin via successive loss of hydroxyalkyl groups, although the mechanism may be more complex than that originally suggested.[28] Triphenyltin is matabolized but not by a hydroxylation route.[29] Aqueous triphenyl tin compounds are degraded by homolytic cleavage of the tin-carbon bonds to Ph_2SnO when exposed to light. Neither Ph_4Sn, monophenyl tin compounds nor inorganic tin are found as products, but a water-soluble organotin polymer is formed $(PhSnO_xH_y)_n$. A similar distribution of products is observed for the decomposition of aqueous di- and mono-phenyl-tin species[29]. It had previously been concluded that Ph_3SnOAc degrades through the di- and mono-compounds to inorganic tin[30] by simple sequential loss of phenyl groups to tin (IV) oxide.

In another study half an application of $(Bu_3Sn)_2O$ disappeared from unsterilized silt and sand loams in approximately 15 and 20 weeks. Small amounts of dibutyl tin derivatives were formed, and carbon dioxide from the butyl groups was evolved. Unextractable tin-containing residues were also formed in the soil. This was ascribed to irreversible adsorption to soil constituents but may have been an organotin polymer, or it could also be tin oxide. The mechanism suggested was hydroxylation[31]. Similar decay to the dibutyl stage in spent anti-fouling coatings has also been reported[32].

It has been shown that residues of $(cC_6H_{11})_3SnOH$ on apples and pears decline by 50% in about three weeks due to photodegradation. A 20-50% reduction may be achieved by washing.

Most studies show that triorganotin compounds are strongly adsorbed onto soil. Over a longer period however studies have not tended to show a build up of tin in the soil to beyond average natural levels in soils. It is likely that environmental leaching over periods of years removes more tin from the soil than is suggested by laboratory studies existing over periods of weeks.

A significant entry of organotin compounds to the aqueous environment arises from the 3000 tonnes used in marine anti-fouling materials. Here release is direct from the ship's hull to the surrounding water and is more environmentally significant in harbours than the open sea. The organotin species adhere to sediment materials rather than remain in solution in water and this matrix appears to be a potential source of any methylation of tin occurring in the environment. Significantly methyl tin compounds have been analysed from harbour waters and sediments.

It can be concluded that the net result of chemical and biological degradation in the environment for alkyltins is a stepwise loss of alkyl groups leading eventually to tin (IV) oxide. Half-lives range from seconds to days and longer in soil. What is uncertain is whether conversion to inorganic tin is always complete, what the timescale for complete conversion is, and whether or not organotins are in effect persistent final decay products.

However studies with $(Bu_3Sn)_2O$ at the 100μg g^{-1} level on soil show no adverse effects on micro-organisms or fertility.

(ii) Mono- and Diorganotin Compounds as Catalysts or Stabilizers

About 20,000 tonnes of mono- and diorganotins are used as heat and light stabilizers for rigid PVC. 2000 tonnes are used as homogenous catalysts for silicone or polyurethane manufacture and transesterification reactions. Here the toxicities of the organotin compounds are incidental to the desired use. The mechanisms of the tin compounds for these uses have been discussed[1,33]. The main routes by which the mono- or diorganotin compounds reach the environment are by leaching from PVC through weathering or solvent action, transport from land burial or transport to atmosphere from incineration of waste products. Leaching from PVC is of most significance where the plastics are used in food contact or as potable water conduits (for the latter use about 3000 tonnes of methyltins were used in the USA in 1981. This low toxicity of methyl derivatives is unusual).

The less toxic mono- and diorganotin compounds degrade along the same pathway as the triorgano products. The concentrations of PVC stabilizers used are at the 0.5-2.0% level. Leaching rates are low, making tin- stabilized PVC acceptable for food contact use, although there have been adverse tissue reactions in medical applications[21]. It is likely that direct biocidal use of triorganotin compounds, although smaller in tonnage terms, leads to a greater introduction of organotin compounds into the natural environment by virtue of the mode of application.

The toxic effects of the lower dialkyl tin compounds are due to their reaction with sulphydryl groups and interference with α-ketoacid oxidation[34]. The higher alkyls are of little toxicity.

Calculations of the amounts of organotin compounds reaching the environment after use have been made for the USA for 1976[35]. A number of detailed compilations of LD_{50} and LC_{50} doses for organotin compounds have been compiled, particularly by the International Tin Research Institute, London[36].

(2) Detection and Transformation of Organotin Compounds in the Environment

A number of organotin compounds have been detected in the environment both close to and remote from the points of application (Table 4). Detection of methyltin derivatives raises the question of their formation. Owing to the use of methyltin compounds in PVC water pipes, the environmental methyltins need not be the product of a natural methylation process in the environment (biomethylation). The levels reported are in the ng dm^{-3} range in air to μg dm^{-3} in urine[37]; ng dm^{-3} levels in water samples were also reported[38]. The latter group detected butyl tins at the μg dm^{-3} level. Two other groups in North America have recently reported the presence of methyl tin compounds in waters, one group at the μg dm^{-3} levels [39,40]. To date mixed methyl-butyl, methyl-phenyl or methyl-octyl tin compounds have not been detected as might be expected if biomethylation of tin were occurring. Possibly complete degradation to inorganic tin has to occur first or even replacement of the other organic groups by methyl.

A number of modern analytical methods have been used to detect low levels of organotin compounds in the environment. Borohydride has been used to convert the organotin species to volatile organotin hydrides. These are trapped and examined later by atomic emission, absorption, mass spectorscopic or GC flame photometric techniques. Butylation to convert to volatile derivatives has also been used.

There is little information on bioconcentration of organotins. There is a report of tin levels of 0.2 to 20 μg g^{-1} in certain marine organisms, suggesting bioconcentration[41]. Tin concentrations in foods after treatment with organotins are about 0.4 to 2.0 μg g^{-1} (surface, apples and pears). Levels on crops do not often exceed 0.5 μg g^{-1} though degradation rates vary. Cows fed with sugar beet containing Ph_3SnOAc at 1 μg g^{-1} produced milk containing 4.0 ng g^{-1} of the acetate. In some countries (e.g. Canada, USA) octyl tin levels of up to 1 μg g^{-1} in food are allowed (from P.V.C. wrapping).[42] There seems little migration of organotins from P.V.C. bottles into liquid foods inside .

There is evidence of bioconcentration by water plants downstream of factories using tin compounds (upstream and downstream concentration ratios were between 170 and 240) and sediments in the region showed increased tin levels downstream of the tin emission[43].

There are now a number of reports in existence of laboratory studies of biomethylation of tin. Some of these are abiotic organometallic studies designed to test the feasibility of possible environmental methylating agents (e.g. methyl cobalamin ($MeCoB_{12}$) or iodomethane). These will not be discussed in detail but are referred to here [44,45]. In general they have tended not clearly to confirm or deny the possibility of environmental tin methylation. $MeCoB_{12}$ is reported to react with divalent tin to give $MeSnCl_3$.[45] Iodomethane produces Me_4Sn with tin powder or with tin (II) salts in the presence of reducing agents.[44]

There are no reports of the incubation of tin (0) or tin (II) or inorganic tin (IV) compounds in sediments to produce Me_4Sn. However a series of incubations leading to methyl tin species has been reported[46]. Various tin (II) and tin (IV) salts were incubated with sediment from Plastic Lake, Ontario, Canada, and the methyl tin products were analysed by conversion to volatile species, followed by GC-AA detection. The trimethyl tin species were found at ng levels.

Incubation of organotin compounds in this system produced Me_4Sn only from Me_3SnCl (at ∿ a $6x10^{-4}$% conversion over 14 days). This production of Me_4Sn could arise by surface catalysed redistribution, sulphide-mediated dismutation (vide infra) or biological methylation.

Butyl and phenyl tin compounds when incubated produce small amounts of trimethyl tin; this would be of larger significance if confirmed at greater yields as it would represent the methylation of the decomposition products of commerical tin compounds to a toxic metabolite. So far the reported yields (about $1 \times 10^{-3}\%$) do not justify concern. Observation of methyl group attachment to inorganic tin compounds in these experiments is important evidence of environmental methylation, though negative results with these sediments have been reported.[47]

The first evidence for methylation of tin (IV) was presented in 1974.[48] Hydrated tin (II) chloride was incubated with a tin-resistant Pseudomonas strain and gave a species having the same fluorescence spectrum as a dimethyl tin control. It was also demonstrated that this strain produced methyl mercury when mercury was added, but much methyl mercury was produced when tin was also present. A transmethylation from tin to mercury was suggested. This work has been extended[49,50], and the results suggest that the organism produces Me_4Sn and methyl stannanes (viz Me_2SnH_2 and Me_3SnH) from incubations with hydrated tin (IV) chloride ($SnCl_45H_2O$). No volatile organotin products were obtained with tin (II) salts or in the absence of tin. The tin (IV) results were explained by biomethylation and reduction. Analysis was by selected ion monitoring with a GC-MS system for the major fragment ions together with calibration chromatograms from standards.

Very recently microorganisms from Chesapeake Bay, U.S.A. have also been shown to transform $SnCl_45H_2O$ to methyl tin compounds[51]. Two methylated tin compounds were observed after borohydride treatment of the culture (viz Me_2SnH_2 and Me_3SnH). Me_4Sn was not detectable in this experiment as it would, if present, have been removed by a purging step before analysis. Identification of products was carried out by GLC retention times and mass spectroscopic techniques.

It has recently been shown that both biologically active and sterile sediments will convert trimethyltin hydroxide to Me_4Sn[52]. Production was greatest in biologically active sediments to which sodium sulphide had been added, but production still occurred at lesser yield in sterile sediments without sulphide. Maximum yield was 6% (high for biomethylation experiments) but there was no evidence that the extra methyl group arose de novo from a methylating agent in the sediment. Various redistribution reactions, some sulphide mediated, could account for the results without biomethylation being invoked. The extra yield in active sediments could be due to more sulphide being present which would assist the redistribution. Some sulphide might be lost on autoclaving the sediments, leading to less Me_4Sn production in sterile sediments.

In view of the laboratory evidence for environmental methylation of tin further studies to detect methyl tin species in the dispersed environment should be made (only North American examples are extant). Particularly in Europe, where less methyl tins are used industrially, would the detection of methyl tins be of great interest. Finally more experiments showing the production of methyl tins in the laboratory from incubation experiments with inorganic tin are urgently required. Present evidence, though compelling, is sparse.

Table 4

Organotin Species Detected in the Environment*

Me_4Sn	$BuSn^{3+}$
Me_3Sn^+	$BuSnH_3$
Me_2Sn^{2+}	Me_3SnH
$MeSn^{3+}$	Me_2SnH_2
Bu_2Sn^{2+}	$MeSnH_3$

*Detected at other than the points of application of commerical organotin compounds.

III Organolead Compounds

(1) Use and Toxicity

The use of organolead compounds as additives to petrol for spark ignition engines has been much debated recently. This is easily the biggest use for organometallic lead compounds. "Knocking" is caused by detonation of the petrol-air mixture rather than smooth combustion, and its occurrence may be removed by adding alkyl-lead compounds to petrol. Lead antiknock additives are Et_4Pb, Me_4Pb and mixed ethyl-methyl alkyl-leads. In a number of countries, amounts used have been reduced recently following legislation; e.g. in the U.S.A. maximum allowable level concentrations in gasoline fell from 2.5 g U.S. gal^{-1} in 1970 to 0.5 g gal^{-1} in 1979; in the U.K. the present level of 0.4 g dm^{-3} will fall to 0.15 g dm^{-3} by 1985. About 200,000 tonnes of lead ethyl anti-knock compounds are presently produced annually throughout the world. This has fallen from 317,000 tonnes in 1974[53].

The toxic effects of Et_4Pb arise by conversion to Et_3Pb^+ in the liver[54]. This is more soluble and attacks the central nervous system. Et_4Pb is readily absorbed through the skin, following which conversion to the trialkyl form takes place, perhaps through initial hydroxylation at a β-carbon position (q.v. organotin compounds). In severe poisoning with Et_4Pb, highest Et_3Pb^+ concentrations are found in the liver. Conversion of Et_4Pb to Et_3Pb^+ in the liver is so rapid that a half-life of minutes for Et_4Pb has been assumed. Elsewhere it is longer. For Et_3Pb^+ and Me_3Pb^+ the half-life in the liver or kidneys of rats is around 40 days for Me_3Pb^+ and 15 days for Et_3Pb^+.

Poisoning with organic lead compounds differs somewhat from that with inorganic lead which generally leads to colic, neurological symptoms and anaemia. The critical organ for organic lead poisoning appears to be the brain. Incipient anaemia is indicated as a first symptom of chronic exposure, and this parallels the case with inorganic lead exposure.

(2) Detection and Transformations in the Environment

Most of the alkyl-lead compounds in petrol are converted to inorganic species during use. These are emitted to the atmosphere, but up to about 10% of the lead released from vehicles (from the exhaust or from evaporation) may be in the organic form. Reactions in aqueous systems will be discussed here though atmospheric levels will be noted. Because most of the organic lead is converted to inorganic lead in use, the environmental problem in the main reduces to the inorganic problem of lead contamination of air, water, food or organisms. The main point of controversy is the extent to which human problems are caused by petroleum - derived lead rather than natural (geological) lead or other industrial inorganic lead emission. The general problem of inorganic lead in the environment will not be covered here but the behaviour of the organometallic species will be discussed.

Organolead species can usually be detected in urban atmospheres but it has been concluded that the lifetimes of the tetraalkyl species in the street air are short[55] and this has been shown for Et_4Pb[56]. Higher concentrations are present near garages and cold-choked vehicles, but in general tetraalkyl lead compounds usually account for about 1-4% of the total airborne lead[57]. These decay to inorganic lead (II) via R_3Pb^+ and R_2Pb^{2+} ions (probably co-ordinated to airborne particulate matter)[112,113], but the major route for decay is photolytic homogeneous reaction in the atmosphere with hydroxyl radical attack being the chief initiation route[58,59]. Photolysis and reaction with ozone also occur. The rate of decay in the atmosphere during daylight has been estimated at up to 21% per hour for Me_4Pb and 88% for Et_4Pb. These pathways do not account for much decomposition at night. Decay routes of this kind seem likely to occur in view of the low concentrations of alkyl-leads found in air despite years of steady emission in certain locations. From various atmospheric samples the proportion of tetraalkyl-lead present has ranged from 0.4% to over 15.0% of the total particulate leads present[60,61] with total tetraalkyl-lead concentrations generally in the range 10 to 200 ng m^{-3},[55,62-65] although a recent case has reported up to 400 ng m^{-3}.[66] In one case it was claimed that up to 62% of the total airborne lead was organic[61]. This is unusual. GLC-MS analysis was used and metal-organic complexes rather than organometallic lead compounds may have been measured. In general greater concentrations of organolead species are found near garages, indoor car parks, road tunnels and in urban air. A monograph in which this question is reviewed has been published recently[19].

Some workers have produced results for "molecular" lead which are rather higher than these but the results may be due to collection problems leading to too high ratios for the real alkyl to total lead content of air. Some of these methods might measure lead organic complexes also[67,68].

Interestingly analysis of urban air in a region where only Et_4Pb (and not Me_4Pb) was used in petrol has shown that methyl-lead species were also present in the atmosphere presumably by chemical rearrangement reactions[56]. The source of the methyl groups is unknown.

Whilst most lead release to the atmosphere is derived from lead in petrol, there is more uncertainty about the origin of the lead contents of soils. Sources of contamination might include agricultural sprays (lead arsenate), lead smelters, power stations (lead in coal) as well as fall out from petrol-derived lead. Recent work with lead isotopes has suggested that, in one location at least, lead from petrol is the major source of the lead found in soils[69].

The proportion of blood lead concentrations due to alkyl-lead in petrol is actively disputed at present. One suggestion is 10%; other groups have suggested higher figures[70]. Reduction of the amount of organolead compounds in gasoline does lead to a reduction in the concentration of lead in air: e.g. in some German cities this concentration was reduced by 60% following the reduction in 1976 of the lead content of petrol from 0.4 g dm^{-3} to 0.15 g dm^{-3} [71]. The extent to which such aerial lead reductions eventually lower blood lead levels is disputed. Further, even the correlation between lead concentrations in blood and the results of behavioural and cognitive tests in humans is disputed. This topic is also of political interest at present.

The decay of organolead compounds in the aqueous environment has been studied[72]. Sunlight, surfaces and certain ions accelerate the decay rate. Suspensions of Et_4Pb and Me_4Pb were quite stable in water in darkness (2% decomposition over 77 days, 16% over 22 days for ethyl and methyl respectively). In the presence of sunlight decay is rapid (99% after 15 days for ethyl, 59% after 22 days for methyl). The rates were catalysed in darkness by copper (II) or iron (II). The products were Et_3Pb^+ and Me_3Pb^+. Only traces of Et_2Pb^{2+} were detected. Absorption of Et_4Pb and Me_4Pb onto silica was complete from aqueous solution and led to decomposition to the trialkyl-lead ions (after 30 days 97% of Et_4Pb and 55% of Me_4Pb had reacted). This suggests that absorption onto sediments promotes the decay of tetraalkyl-lead species in the environment. Some groups have reported little absorption but others have detected these compounds on sediments[73]. In darkness solutions of Me_3PbCl, Et_3PbCl and nBu_3PbCl are stable.[72] Only Me_3PbCl shows decay (1% reaction after 220 days). Metal cations had no effect. Sulphide anion promoted the production of R_4Pb. Sunlight increased the decomposition rate; in 15 days there was 4% loss of Me_3Pb^+, 25% loss of nBu_3Pb^+ and 99% loss of Et_3Pb^+. The main detectable product was inorganic lead. Presumably R_4Pb also was formed. Again silica absorbed the organolead cations completely from water and the breakdown rates were promoted slightly.

Dialkyl-lead compounds disproportionate slowly in darkness (after 30 days 10% of Me_2Pb^{2+}, 6% of Et_2Pb^{2+} and 4% of nBu_2Pb^{2+} had reacted). Trialkyl-lead products and lead (II) were formed.

In sunlight after 40 days 70% of nBu_2PbCl_2, 25% of Et_2PbCl_2, and 5% of Me_2PbCl_2 had disproportionated. This and other work suggests that alkyl-lead compounds emitted from vehicles and deposited in waterways undergo fairly rapid decomposition in the presence of light, surfaces or ions. In theory any inorganic lead formed might undergo bio-methylation to Me_4Pb. This is considered later. Decomposition of alkyl-lead cations in water has also been studied by other groups with rather similar conclusions[74,75]. In sea water it has been suggested that Et_4Pb would lie on the seabed as a separate phase, slowly dissolving into seawater. Some would evaporate, but most would form Et_3Pb^+.

Et_4Pb has been analysed at a 30 $\mu g\ g^{-1}$ level in mussels near a sunken ship which had been carrying Et_4Pb[76]. These results suggest that alkyl-lead compounds are not quickly metabolized by organisms and may remain in the organic form in tissue for some time. The occurrence of tetraalkyl-lead in aquatic biota is significant because of the possibility of food chain effects.[77] However in one example only one out of 50 Canadian fish was observed to contain Me_4Pb, (0.25 μg g^{-1}) and the source was unknown[78].

Where Me_4Pb is present it may be accumulated by fish either through water or food[79,80]. One work suggests a rate of 0.4 μg g^{-1} to 2.5 μg g^{-1} daily from a concentration of Me_4Pb in water of 3.5 μg dm^{-3}, giving a daily accumulation factor from water of 100-700. Most Me_4Pb is accumulated in fatty tissue and the concentration factor in lipids in the intestine was calculated at 16,000 from water containing 25 μg dm^{-3} of Me_4Pb. The gills, air bladder and liver also accumulate Me_4Pb. The half lives for loss of Me_4Pb from intestinal fat and skin were 30 and 45 hours to lead-free water. This presumably occurs by metabolism to Me_3Pb^+ derivatives, the basic toxic substance[80,81]. Accumulation of alkyl lead has been found in cod, lobster and mackerel tissue where levels of tetraalkyl-lead of between 0.1 and 4.79 μg g^{-1} were found. In lobster digestive gland the tetraalkyl-lead was 81% of the total lead. Percentages of organic lead in various matrices range from 9.5 to 89.7 (flounder meal). The analytical method was extraction with benzene/aqueous EDTA solution followed by flameless AA[82].

In algae Me_4Pb accumulates in the cytoplasm[79].

An analysis of 107 fish showed 17 samples containing tetraalkyl-lead compounds. In this study water, vegetation, algae, weeds and sediments showed no organic lead to be present. Some of the fish were caught in locations far from likely contamination use (viz parts of Ontario, Canada) and biomethylation was mentioned as a possible cause for the presence of tetraalkyl-lead compounds. In general the alkyl-leads were less than ten per cent of the total lead in the fish[83,84].

Much analytical development work for alkyl-lead analysis has been carried out. Recent methods used include GLC-AA with electrothermal atomization[56], GLC with microwave plasma detection[85] and flameless AA with heated graphite atomization[86]. Other analyses have also been performed by similar methods[87-91].

Tetraalkyl- or other alkyl-lead species have been measured frequently in the environment and their origin is usually ascribed to evaporation of unused petrol or to incomplete combustion of the organoleads in petrol but sometimes bio-methylation of inorganic lead is cited.

Detection of methyl-lead species in the environment is not, in itself, evidence of biomethylation and some of the sediments used in lead methylation studies already contained tetramethyl-lead[73,92]. There are a number of reports that inorganic lead may be methylated in the environment. A mixture of Great Lakes, Canada, water and sediments with nutrients has produced Me_4Pb without any addition of lead in the laboratory; these sediments already contained lead. Addition of Me_3PbOAc greatly increased the amounts either through disporportionation or biomethylation. Addition of some inorganic lead (II) salts also caused an increase in the amount of Me_4Pb present. Using pure species of various bacteria up to 6% of Me_3Pb^+ present was converted to the tetramethyl form in one week. Inorganic lead was not converted under these conditions. It is conceivable that some of these observations could be due to Lewis acid displacement of weakly bound pre-existing Me_4Pb, as lead was already present in the sediment. It was not demonstrated that the lead added was the same as the lead later analysed as Me_4Pb[92]. This possibility might be unlikely but it should be borne in mind. Analysis was by GC-AA.

Me_3Pb^+ salts can also be converted to Me_4Pb by chemical disproportionation. One route involves sulphide ions or hydrogen sulphide in an analogous route to that existing for tin or mecury[93]. Sulphide mediation may be of general importance in the conversion of partially substituted alkyl metals in the environment to the fully methylated species. There is agreement that Me_3Pb^+ salts are converted to Me_4Pb in the environment, but there is debate about the proportion of the conversion arising from redistribution of the existing methyl groups (disproportionation) and the proportion arising from biological methylation.[93-95]. This latter has been calculated to be up to 20% of the whole methylation[96]. However different experimental conditions between sterile and biological experiments might have changed the rate of the disproportionation mechanism on which the calculation was based. One group has suggested a biological methylation occurring ten times faster than the sulphide promoted route[97], but other workers have not invoked a biological component. One group has found that the conversion was up to 4%[98], a result in agreement with previous work by other groups using sediments from widely differing locations[92,95]. It has been suggested that where the concentration of lead is insufficient to stop growth of micro-organisms, the proportion of

Me_3Pb^+ methylated biologically is over 80% with less than 20% converted chemically[73]. In a different system a biological methylation of 50-76% of the total was claimed[99] but the yield was only 0.009%. Et_3PbCl has been methylated as well as disproportionated in a micro-organism culture; 13% of the methylated derivative was found[96], but another group found no methyl group was added to the Et_3Pb moiety.

There are other examples of the conversion of lead (II) salts to Me_4Pb. In a series of water samples seeded with micro-organisms from an aquarium, incubation of lead (II) acetate produced Me_4Pb (N.B. mercuric acetate may produce methyl mercury by methylation from decomposition of the acetate methyl group alone, but under more extreme conditions[101]). It has also been reported that lead (II) salts added to St Lawrence River sediment will produce Me_4Pb. Two out of three sediment sites produced this from lead (II) nitrate. No control experiments were reported and analysis was by GLC retention times[97]. Results of incubations with anoxic British Columbia sediments containing added lead (II) nitrate suggested a 0.03% conversion to Me_4Pb[102]. Control experiments were reported and identification of the product was by GLC-MS method. Although these sediments also contained lead, Me_4Pb was not observed in unspiked samples suggesting that the results were not due to displacement of pre-existing organic lead. In this study lead (II) acetate did not methylate. However, a more recent report by these workers has placed doubt on the existence of lead (II) methylation here[98]. Using ^{14}C-labelled methyl donors (L-serine, methanol, MeCoB12 and D-glucose) no ^{14}C-methyl was found in the Me_4Pb arising from incubation experiments with Me_3Pb^+, suggesting chemical methylation only[103,104]. Some groups, however, have been unable to detect lead (II) methylation in micro-organism or sediment media[93,95,103,104]. There is indirect evidence of lead methylation based on reverse air movement projections. An enhancement of the alkyl to total lead ratios[86] normally found in airborne particulate materials was observed. It was concluded that in certain Cumbrian, U.K., intertidal sediments conversion of inorganic lead to organic lead was taking place. Analysis was by absorption in iodine monochloride, extraction into acid and AA.
Much abiotic organometallic work has been carried out recently to simulate lead methylation in the environment. The principal conclusions will be mentioned here where they are relevant to the environment.

Organolead compounds have not been detected from the reaction of $MeCoB_{12}$ with lead (II) salts[93,105-108]. The naturally occurring methylating agent iodomethane (MeI) is variously reported to a) react and b) not react with lead (II) salts in aqueous media to produce low yields of Me_4Pb by oxidative addition. There is active controversy at this time concerning the reaction.

Abiotic chemical alkylation of lead (II) in water to give Me_4Pb is not inherently impossible. It has been carried out with boron alkyls[109,110] by methyl carbanion alkylation (q.v. $MeCoB_{12}$). There is, then, little theoretical reason why lead (II) might not be alkylated under environmental conditions; although monomethyl-lead species are unstable, if the rate of further methylation to the more stable R_2Pb^{2+} or R_3Pb^{+} derivatives is greater than decomposition, then Me_4Pb may be formed - as was observed for the boron alkylating agents. It is also possible that monomethyl-lead species might be stabilized in the environment by co-ordination to natural ligands.

The question of lead (II) methylation remains open. There is a need for isotopic work which, in a methylating system, would clearly demonstrate that the lead added is that which is methylated. Ideally analysis should be in an absolute method (eg GLC-MS).

III Organomercury Compounds

(1) Uses and Toxicity

Much has been said about the environmental role of organomercury compounds and much research is still being carried out. A number of detailed accounts exist and the discussion that will appear here will be closely defined. As for lead, increases in total metal concentrations today compared to the pristine environment are not solely caused by use of organomercury products. The main use for mercury has been as the metal as the cathode in chlor-alkali cells and in electrical apparatus, and most mercury introduced to the environment has probably come from these sources. Only organomercury concentrations and uses will be discussed.

Restrictions on organomercury compounds have reduced the quantities released to the environment. Uses of organomercury compounds are covered in publications by the Organization for Economic Cooperation and Development [111] and the International Atomic Energy Agency [112], but these are rather historic in terms of present use. Use of alkyl mercury compounds has declined and other organomercurials are generally being further restricted in use.

The chief toxic effects of inorganic mercury poisoning are tremor, psychological disturbance, gingivitis and occasionally proteinuria. Acute doses of mercury (II) lead to kidney injury and perhaps death. The symptoms of methyl mercury poisoning are well known. The critical organ is the brain and penetration of the blood-brain barrier leads to sensory disturbance, tremor, ataxia, constriction of the visual fields and impaired hearing [113]. There is often a long latent period after exposure and the effects are frequently irreversible. At Minamata pre-natal exposure occurred, giving symptoms in the infants of cerebral palsy characterized by mental retardation and motor disturbance. The mothers may or may not show symptoms.

Elimination of methyl mercury is slow, and this species produces more mercury in the brain than intake of other forms of mercury. Excretion eventually takes place to about fifty per cent inorganic form. It is the symptomless build-up coupled with the acute effects which is so serious. The half life in the human body for methyl mercury is about 70 days, compared to 4 or 5 for mercury (II) and 10 for $MeOCH_2CH_2Hg^+$ in the rat [18,114].

Excretion rates of methyl mercury are slow. After 10 days for chicks 20% of methyl mercury, 60% of inorganic mercury (II), 80% of phenyl and 90% of $MeOCH_2CH_2Hg^+$ had been eliminated. Half lifes for methyl mercury in chickens are 70 days, osprey 2-3 months, mallard drakes 84 days; similar to man [115]. From fish half lifes of up to 1,000 days have been measured but as low as 8-23 days in goldfish [116].

A daily intake of methyl mercury of 0.3mg for the average human can lead to the appearance of toxic symptoms. Acceptable weekly intake levels of 0.3mg of total mercury were defined by the World Health Organization and the Food and Agriculture Organization.

This leads to allowable maximum levels of about 0.5 to 1.0 μg g^{-1} in fish and shellfish[11,17,65,117].

It is estimated that the lowest whole blood mercury level assumed capable of producing neurological symptoms is about 0.2 μg g^{-1}. To prevent this level being reached limits for mercury concentration in foods have been set in various countries (e.g. for fish these are 1.0 (Sweden), 0.5 (Canada), 0.5 (U.S.A.), 1.0 (Finland) and 0.4 (Japan) - all in μg g^{-1})[117]. Phenyl and alkoxyalkyl mercury compounds break down to organic materials (e.g. ethene) and inorganic mercury and with their shorter half lives are more akin to inorganic mercury in their effects. For these compounds the kidney seems to be the critical organ[118]. Methyl mercury also decays in soils or sediments to mercury (0) but again at a slower rate[119-121] than phenyl or alkoxyalkyl derivatives[122,123].

In cases of prolonged exposure toxicity becomes chronic at lower concentration, and metal levels in the tissues can be higher than those found in cases of acute poisoning[124]. A number of studies of the toxicity of organomercury compounds have been made[3,17,18]. In one case fish-eating populations near the Agano river in Japan who consumed 300-1500g of fish daily, with the methyl mercury level in the fish at 3-4 μg g^{-1}, suffered poisoning. The rate of methyl mercury uptake by fish from water may be 10-100 times faster than for mercury (II) ion but its elimination is slower. Hence there is particular concern about methyl mercury levels in fish which is a main pathway exposing man to mercury.

Table 5

Uses of Organomercury Compounds

Compound	Use	Notes
MeHgX	Seed dressings (fungicides)	Banned Sweden 1966.
EtHgX		USA 1970, Canada 1970 etc.
RHgX (R = Ac, py)	Catalysts for urethane, vinyl acetate production Slimicides	Little used.
PhHgX	Seed dressings (fungicides) Bactericides, Slimicides	Banned as slimi cide USA 1970, banned for rice Japan 1968 etc.
pMePhHgX	Spermicide	
RPhHgX (Substituted phenyl)		
$MeOCH_2CH_2HgX$	Seed dressings (fungicides)	Banned Japan 1968
$EtOCH_2CH_2HgX$		
Thiomersal (EtHg derivative)	Antiseptic	Reducing in usage
Mercurochrome (organomercury fluorescein derivative)	Antiseptic	Reducing in usage
Mersalyl (methoxyalkyl Hg derivative)	Diuretic	Reducing in usage
Chlormerodrin (alkoxyalkyl Hg derivative)	Diuretic	Reducing in usage

X = inorganic or organic anion

(2) Detection and Transformations in the Environment

Methyl mercury concentrations have been measured in many environmental matrices during the past fifteen years. Much data is available in the work edited by Nriagu[3]. The emphasis here is to compare measured methyl mercury levels with calculated pre-man environmental levels, to compare the ratios of organic to inorganic mercury in various media and also to assess bioconcentration effects in various food nets. Reported methyl mercury levels are a measurement of an equilibrium between methylation and demethylation. It should not be assumed that any methyl mercury detected is necessarily a man-made pollution problem; there is evidence to show that levels in certain marine species are much the same today as they were centuries ago. Environmental levels of organomercury species are available[3,6,11].

The organomercury content for fish and freshwater biota has been investigated in detail. Most freshwater fish have analysable levels of mercury in their tissues and usually more than 80% is in the methyl form. All forms of mercury seem to be accumulated by fish from water and food and methyl mercury is absorbed faster than inorganic mercury and retained longer. In other aquatic species the percentage of the methyl form seems to vary between about 50% and 80%. The concentration of methyl mercury in fish tends to be proportional to length and age, and accumulation is usually proportional to mercury concentration in water and food. Accumulation is correlated with temperature to a certain value for each species which is connected with oxygen consumption rates.

Methyl mercury is absorbed by fish from bottom and suspended sediments and food and possibly as the dissolved species from water. There may be no actual methylation in fish (though in vitro methylation from homogenates has been demonstrated[125,126]), but demethylation occurs slowly and so methyl mercury accumulates.

Freshwater biota, even from remote areas can accumulate detectable quantities of mercury from natural and crustal sources. Fish from such areas still contain ng g^{-1} concentrations, of which up to 90% is in the methyl form[127]. Acid mobilization of mercury in remote areas and deposition of mercury from the atmosphere may ultimately be man-derived as these processes provide more mercury for methylation.

Data on dissolved methyl mercury may not really refer to actual dissolved species, rather to mercury complexed to suspended particulate matter. A sample of interstitial water has been reported to contain 1.4 $\mu g\ dm^{-3}$ of methyl mercury. Dissolved mercury has been reported at 0.01 - 0.03 $\mu g\ dm^{-3}$ from one source, while some Canadian lakes contained 0.5 - 1.7 $ng\ g^{-1}$.[128-130]

The proportion of mercury in the methyl form in birds varies considerably depending on the mercury source and may vary between 10 & 100%.[1]

The best indicator of methyl mercury exposure in humans are hair, urine or blood. The average methyl mercury level in hair in Japan was reported to be 2.07 $\mu g\ g^{-1}$; people living in a mercury polluted district had 6.69 $\mu g\ g^{-1}$, heavy fish eaters from Oyabe had 11.21 $\mu g\ g^{-1}$ and tuna fishermen had 12.82 $\mu g\ g^{-1}$. Pre-industrial examples of this latter group may have had similar levels.[131]

As most mercury in fish is in the methyl form, some total mercury levels will be discussed. Marine fish having longer live spans have the highest levels of mercury. These are from about 0.1 to 0.3 $\mu g\ g^{-1}$ dry weight for small pelagic fish whereas larger species such as tuna and swordfish are between 0.5 and 1.5 $\mu g\ g^{-1}$, mostly as methyl mercury. Predators accumulate higher levels of methyl mercury than fish or other species lower in the food net.[132] There seems little difference in the percentage of mercury in the methyl form whether or not the fish comes from a contaminated location. Apart from methyl mercury in fish, there is a report of ethyl mercury being found in fish downstream of phenylmercury effluent.[133]

Maximum background levels of mercury in freshwater fish are about 0.2 $\mu g\ g^{-1}$ (muscle) although more than 1.0 $\mu g\ g^{-1}$ can be found from geological mercury sources alone. The minimum amount for freshwater fish is about 0.035 $\mu g\ g^{-1}$.[127] Fish from polluted rivers show higher total mercury levels; e.g. in 1965 fish from the Agano River (Niigata, Japan) contained an average of 2.34 $\mu g\ g^{-1}$ of mercury. Interestingly this has decreased in recent years (to 0.60 $\mu g\ g^{-1}$ in 1968, and to 0.32 in 1971).[131] A similar decrease in methyl mercury levels in sediments from the River Mersey, U.K., has been observed recently.[134-137]

Analysis of methyl mercury is usually carried out by the Westoo technique or variations. Samples are extracted with toluene or benzene, and methyl mercury is extracted to an aqueous phase followed by reextraction into benzene or toluene. Analysis is usually by electron capture GLC[135-140]

Biota in oligotrophic lakes often show higher methyl mercury levels than those in otherwise similar eutrophic lakes. This may be due to dilution in the higher biomass of the latter, and also to the higher pH favouring volatile dimethyl mercury formation. In eutrophic lakes the sediments are largely reduced, forming mercuric sulphide which is little available for methylation.[141] High sulphide levels tend to promote conversion of methyl to dimethyl mercury[142], viz $2CH_3Hg^+ + S^{2-} \longrightarrow (CH_3Hg)_2S \longrightarrow (CH_3)_2\ Hg + HgS$
Oligotrophic lakes subject to acidification (e.g. acid rainfall) have less productivity, lower biomass and greater availability of mercury. At low pH methyl mercury is favoured and this is absorbed by the reduced biomass - hence the higher concentrations. Reduced complexation to sediments also leads to greater bioavailability, which may, however, be opposed by lower microbial activity.[143] Acid precipitation and methylation and mobilization have been reviewed recently.[144]

Several observations suggested that inorganic mercury could be methylated in the environment; (i) most of the mercury present in fish is present in the methyl form[145], (ii) inorganic mercury when added to aquarium sediments is partly converted to methyl mercury[146] and (iii) $MeCoB_{12}$-utilizing methanogenic bacteria can methylate mercury in sediments.[147]

It has now been demonstrated on many occasions that inorganic mercury added to sediments may be converted to methyl mercury. In addition various pure bacterial strains with the capacity for mercury methylation have been isolated.[148-150] However several authors report an inability to methylate mercury by various systems.[103,104,151,152] This must be a property of the particular system or strain studied; mercury methylation in general terms is firmly established, and the main parameters seem to be as follows:

Both methylation and demethylation[153-155] occur in sediments and measurements of methyl mercury are normally measurements of an equilibrium process. Aerobic and anaerobic methylation and demethylation occur[156-163]. Methylation is mainly to monomethyl mercury under normal conditions and demethylation is to methane and mercury (0). Workers have on occasions found faster net methylation rates under aerobic[148,164,165] or anaerobic[166] conditions. Factors which influence the rate of methylation include total mercury concentration, silt and organic content of a sediment, pH, Eh, temperature, concentration of methanogenic bacteria, sulphide content and complexation. It has been concluded that maximum rates of methylation are found in the oxidising anaerobic zone where redox potential (Eh) ranges from about -100 to +150mV[135]. There is an environmental mercury cycle by which mercury compounds may be inter-converted. This explains why methyl mercury was found in fish downstream from pulp mills (phenyl mercury in the effluent) and chlor-alkali plants (inorganic mercury). The mercury cycle is discussed in detail in Reference 1. It seems that methyl mercury is preferentially formed under neutral and acidic conditions and dimethyl mercury under basic conditions[167]. The presence of sulphide ion in natural systems reduces methylation by formation of the largely unavailable mercuric sulphide. Any methyl mercury present may be converted at higher sulphide levels to the dimethyl form by formation of an organo mercury sulphide intermediate (see above)[168,169]. Mercuric sulphide may be microbially converted to the sulphate so sulphide formation is not irreversible. Methyl mercury does not seem to build up in sediments beyond about 1.5% compared to total mercury[135]. This is a rough equilibrium level between formation and removal by various processes. Percentage levels in fish and other biota may be much higher. Methyl mercury is also produced in the water column[170] and in soil both biologically[156,158] and abiotically[171,172] MeHgSMe has been found in shellfish[173,174]. In soils dimethyl mercury may be an important product[175-177]. It is not established that fish themselves can methylate mercury, although in vitro extracts may do so. It may be that absorption of methyl mercury from outside is the main cause of methyl mercury concentration in fish. Net methylation rates from sediments and rivers have been measured and range from 17 to 690 ng m^{-2} day^{-1} (river),[178] 15-40 ng g^{-1} day^{-1} (sediment),[146]

15 ng g^{-1} day^{-1} (sediment)[154] and 137 ng g^{-1} day^{-1} (marsh sediment)[179]. A saline environment produces less methyl mercury than equivalent non-saline regions, suggesting methylation may be a methyl carbanion transfer[180] (i.e. invoking Me CoB_{12}). Further discussions on mercury methylation may be found in several recent reviews[5,164,181].

IV Other Organometallic Compounds

(1) Use and Toxicity

Release of organoarsenic compounds to the environment is declining. The main organometallic uses are as pesticides and herbicides. Pentavalent arsenic compounds, particularly the aryls, have in the past been much used as bactericides and pest control agents.

A number of derivatives of phenyl arsonic acid ($PhAsO(OH)_2$) have bactericidal, biocidal or pharmaceutical properties. Arsenilic acid ($pNH_2PhAsO(OH)_2$) and its sodium salt are listed in the British Pharmacopoeia (Veterinary)[16]. Agricultural or veterinary uses of compounds are less controlled than medical use and there is greater potential for release to the environment and for toxicity problems. Inorganic arsenic may be methylated in the natural environment. The input of organoarsenic compounds into the environment by man may be insignificant in comparison with natural arsenic movements, particularly as some of the compounds are common to both systems. In fact the marine arsenic cycle appears to be a detoxification process by the organisms concerned. This is discussed later. Where arsenic compounds are found their origin may be from anthropogenic or natural processes, or, in certain locations, from both.

The use of arsenic compounds in pesticides has been the subject of a monograph[182]. Although of minor importance now in many countries, the organoarsenic drugs are still available. For herbicidal use the organoarsenicals and other pesticides mainly replaced inorganic arsenic pesticides or defoliants (e.g. lead, calcium arsenate, sodium arsenite, arsenic trioxide). High rates of application of the inorganic compounds were required and this caused arsenic build-up in the soil, severely reducing plant growth. The organic arsenical herbicides that have been in common use are shown in Table 6. They are methyl arsonic and

Table 6 Uses for Organoarsenic Compounds

Compound	Use
$Me_2AsOONa$	Herbicide
$MeAsO(ONa)_2$	Herbicide
pRPhAsO(OH)ONa, where R = NH_2, $NHCH_2CONH_2$ or $NHCONH_2$	Trypanosomiasis Drug
pRPhAs< (ring: S—CH_2, S—$CHCH_2OH$)	Trypanosomiasis Drug
$R^1R^2PhAs = AsPhR^1R^2$ where R^1 = pOH, R^2 = oNH_2	Antisyphilis Drug
$pNO_2PhAsO(OH)_2$	Animal feed additive
$R^1R^2PhAsO(OH)_2$, where R^1 = pOH, R^2 = mNO_2	Animal feed additive

Table 7 Some Organoarsenic Concentrations*

Matrix	MeAs (ng g^{-1})	Me_2As (ng g^{-1})	Me_nAs % Total As
Pond	7.4		50
Saline lake	0.7	6.75	10-90
Lake water	0.2	0.25	20
Sea water	0.004	0.02	10-90
California (Sea)	0.1 - 0.2	1	
Marine Biota#	0 - 0.05	0 - 0.24	10-90
Kelp			4
Macroscopic algae	0.4 - 8.0 x 10^3		
Shark muscle			40-80
Shale process waters	200		
Fly ash slurry	24	109	

* Relatively few organoarsenic levels have been determined for environmental matrices and more data are required before these values can be considered as typical

\# Diatoms, Coccolithophorids, Dinoflagellates, Prasmophyceae

dimethyl arsenic (cacodylic) acids or salts. The most extensive use in the USA has been for control of annual grass weeds in cotton. Since 1945 organoarsenic compounds have been shown to have both therapeutic and growth properties as feed additives for poultry and swine. The toxicity of arsenic compounds is different from those of the heavy metals. For arsenic the toxicity to the rate declines with arsenite > arsenate > methyl arsenate = dimethyl arsenate (cacodylate). The toxic dose increases by about fifty between arsenite and the methyl acid salts[183]. The toxic effects appear to be caused by binding to sulphydryl lipid groups by trivalent arsenic, and pentavalent arsenic appears to be reduced to the trivalent form. Lower doses produce liver and kidney damage, while acute administration leads to enteritis and death. Arsine poisoning results in anaemia and renal damage. The possible role of arsenic as a cause of cancer is still not clear[184].

(2) <u>Detection and Formation in the Environment</u>

As a ubiquitous element in water, air or soil and living tissue, the detection of arsenic in a sample is not in itself indicative of man-derived contamination. The existence of a natural biomethylation of arsenic in the natural environment reinforces this position, particularly as it increases transport possibilities for arsenic.[185] Arsenic concentrations are given in Table 7. Natural processes, when taken together with the use of both organic and inorganic forms of arsenic by man, can make it difficult to assess the true cause of a particular arsenic level.

The chemistry of arsenic in the marine environment is complex and fascinating. Although arsenate is the dominant species in waters of the photic zone, arsenite, $MeAsO(OH)_2$ and Me_2AsOOH are found in significant concentration[186]. There is a correlation between the methyl arsenic species and biological productivity (e.g. chlorophyll concentration and carbon uptake) which suggests the importance of biological activity in arsenic speciation. The methyl arsenicals appear to be formed from and return to the main form, arsenate, and a marine arsenic cycle has been derived. The formation and cycling of these species occurs with natural levels of arsenic and is not a result of man's activities.

The same methyl arsenic species are also observed in macroscopic algae. A number of other methylarsenic compounds are formed in the marine environment. Biosynthetic organoarsenical compounds are shown in Figure 1, together with the arsenic cycle.

$MeAsO(OH)_2$ and Me_2AsOOH can also be found in freshwater rivers and lakes. Reduction of arsenate to arsenite in seawater by various species has been demonstrated and the reverse process has been known for some time. Methylation seems to proceed initially by reduction of arsenate as demonstrated in laboratory systems[187-190].

Below the photic zone the methylated arsenicals rapidly decrease to levels below detection limits (<pg g^{-1}). Arsenic is concentrated by aquatic organisms but not a great deal is known about accumulation ratios for organic arsenic compounds. Algae accumulate more arsenic than fish, with crustacae accumulating intermediate amounts. The chemistry of arsenic within plant and other aquatic species has been discussed above[191]. Arsenobetaine ($Me_3\overset{+}{A}sCH_2CO\bar{O}$) has been found as 80% of the arsenic present in shark muscle and this derivative is thought to be the major form of arsenic in higher animals. For shark liver and lipid 40% and 45% respectively were in this form[192]. Origin of the methylarsenic compounds is in the substitution of arsenate for phosphate in metabolism in the oceans at low phosphate levels. By conversion to organic forms of arsenic the algae detoxifies the arsenate originally taken in and a critical point is an organo arsenic phospholipid.[193-194] The final sequence is postulated to be degradation to the arsonium lactate, $Me_3\overset{+}{A}sCH_2CHOHCO\bar{O}$, which probably degrades further to Me_2AsOOH salts[194]. Reductive methylation of arsenate to Me_3As and formation of the lactate is also believed to take place in terrestrial plant roots and may account for the arsenic levels found in various plants and trees. We can therefore consider arsenic methylation as a detoxification process, but as arsenic serves no known biological function[184] in living organisms, arsenic metabolism must still be viewed as a contamination rather than as part of an inherent metabolic process. However, such contamination is usually a consequence of the natural cycling of this element and it is not particularly man-derived.

Figure 1

The Marine Arsenic Cycle

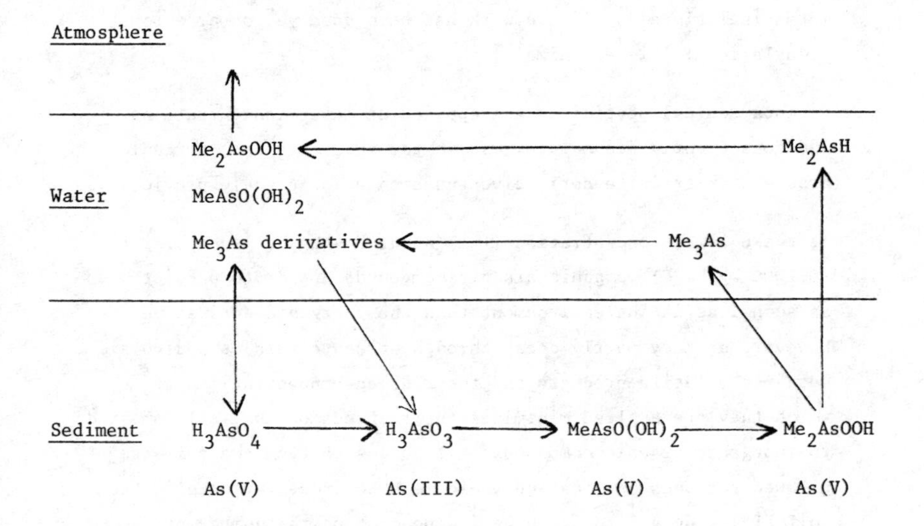

Organoarsenic Compounds Found in the Environment

$MeAsO(OH)_2$	Me_2AsH
Me_2AsOOH	$Me_3\overset{+}{A}sCH_2CH_2OH$
Me_3AsO	$Me_3\overset{+}{A}sCH_2CO\bar{O}$
Me_3As	$PhAsO(OH)_2$

$Me_3\overset{+}{A}sCH_2CH(CO\bar{O})OPO_2OCCH_2CH(OCOR)CH_2OCOR^1$

Hence, most of the levels of organoarsenic compounds shown in Table 7, though superfluous to the organisms concerned, are natural levels. It should be noted that most of the inorganic arsenic ingested by humans is excreted in the urine as the methyl arsenic compounds[195,191]. Methylation is presumably by intestinal flora but little work has been done yet on arsenic methylation in higher animals.

Organoarsenical pesticides are applied at lower concentrations and consequently accumulate less in the soil than the inorganic arsenate or arsenite derivatives used on a large scale previously.

At these lower concentrations (viz about 2-4 kg ha^{-1} compared to 10-1000 kg ha^{-1}) organic arsenic compounds appear to persist as such less in the environment than the inorganic derivatives. However, as they partly decay through arsenate this is equivocal. Decay to volatile products and the reduced concentrations at which they are applied minimizes their effect on the soil compared to inorganic arsenic compounds. If no losses from the soil are allowed for such application would lead to excess arsenic in the soil of 1.4 μg g^{-1} to 4.6 μg g^{-1} dependent on the number of applications and the half life[197]. Losses due to methylation and reductive volatilization, erosion and crop removal may however reach 40% - 60% of applied arsenic.

As a consequence of agricultural use, soil absorption studies of organoarsenics show absorption is rapid,and further changes to less soluble forms then occur. Applied arsenate and dimethyl arsenate (cacodylate) become gradually less extractable over six months after addition. Although the arsenic-carbon bond is stable in plants, microbiological oxidation takes place in soils producing up to ten percent yields of carbon dioxide over 30 days, together with arsenate. A series of experiments have shown that over 24 weeks under aerobic conditions about 41% of applied Me_2AsOOH is oxidised to carbon dioxide and arsenate while 35% is lost from the soil as a volatile organoarsenic compound[198]. Under anaerobic conditions there is no production of carbon dioxide but 61% was volatilized. These appear to be biomethylation processes by fungi and methanogenic bacteria respectively.

Biomethylation also takes place for phenyl arsenic compounds in the natural environment. Rate of oxidation of organoarsenicals to arsenate in various soils seems to be between two and ten per cent per month. This leads to estimates of 0.78 and 0.28 respectively, remaining one year after application at unit concentration, giving half lives of 34.3 and 6.6 months in each case[125] ignoring volatilization. Both organic and inorganic forms of arsenic in plants seem to absorb and translocate in relatively small amounts.[199] Experiments with MeAs(O)(OH)ONa over a six year period showed no arsenic residues in the cottonseed[200]. Most of the arsenic losses were not to the plants but by methylation and adsorption in situ to soil. Greater absorption may occur in roots. In general, little arsenic is found in plants after arsenic treatment. To achieve significant levels, $Me_2AsO(ONa)_2$ at 35 kg ha^{-1} had to be applied to the soil, ten times the normal rate. Most arsenic is volatilized, eroded or retained as insoluble forms with aluminium or iron compounds in the soil[197, 2.1].

At normal rates of use, organoarsenic compounds fed to animals do accumulate in tissues but to amounts less than the elevation of arsenic in the animal food over natural levels. In poultry and swine a high proportion of ingested arsenic is rapidly excreted unchanged[202]. The excreted arsenic species may then be subject to the reactions in soils described above. There is no evidence for conversion to inorganic forms of arsenic within the animal. $MeAsO(OH)_2$ and $Me_2AsO(OH)$ have been found in shells and human urine but not in rainwater[203,204].

Use of organic arsenicals in animal feed (permissible levels, 50-375 μg g^{-1}) does lead to arsenic residues in tissues. Feeding the approved levels gives 1-2 μg g^{-1} in chicken liver; feeding at ten times the approved level gives 6-7 μg g^{-1} (permissible level, 0.5-2.0 μg g^{-1}). As noted, these levels rapidly decline after arsenic-containing feed is withdrawn[202]. Feeding of arsenilic acid ($pNH_2PhAsO(OH)_2$) at the 273 μg g^{-1} level (nearly three times that approved for poultry or swine) produced arsenic levels of 29.2 (liver), 23.6 (kidney) and 1.2 μg g^{-1} (muscle) respectively. Original levels were less than 0.01 μg g^{-1}.

The wholeblood level increased from 0.01 to 0.54 µg g^{-1}. All weights are dry weight.[202] The depletion period was examined following withdrawal of arsenic and it was found that after six days liver arsenic levels had fallen from 29.2 to 5.0 µg g^{-1}. Most arsenic compounds are excreted unchanged from animals. The use of arsenic-containing animal waste as fertilizer has been considered from the point of view of increase in the arsenic burden of the soil; fortunately, it would take more than 2,000 tonnes of waste containing 15-30 µg g^{-1} arsenic applied each year per acre to raise the soil arsenic to the 500 µg g^{-1} level found in some arsenic-treated soils.

Analysis of arsenic in environmental matrices is by a number of methods.[127,213,205-208] Conversion to arsine derivatives by zinc and hydrochloric acid, electrolysis or by borohydride is commonly used to speciate arsenite, arsenate, $MeAsO(OH)_2$ Me_2AsOOH. The latter evolve methyl arsines. The volatile arsines are trapped and measured by AA or atomic emission. Photometric measurement of arsenic complexes is also used.[209-211] Historically, methylation of arsenic (III) compounds to Me_3As by cultures of the mould S. brevicaulis by Challenger was the first demonstration of biological methylation of a metalloidal element. $MeAsO(ONa)_2$ and $Me_2AsO(ONa)$ also gave Me_3As, whereas alkyl arsonic acids ($RAsO(OH)_2$, $R_2AsO(OH)$) produced the appropriate methyl alkyl arsine, confirming the supply of methyl by the mould. The methylations took place in well aerated mould cultures which nevertheless exhibited strong reducing action. The mechanism appears to operate by methyl carbonium ion transfer from S-adenosyl methionine (SAM) in the mould, to a lone electron pair on arsenic(III), (i.e. an oxidative addition). Further reduction and subsequent methylation may occur as appropriate. This explanation was put forward by Challenger in two reviews.[212,213] It is possible that methylation of Group IV divalent elements (tin and lead) might also involve methyl carbonium transfer (via SAM) as leading to the observed tetravalent methyl compound. Soil organisms have also been shown to produce Me_3As.[214] Anaerobic synthesis of Me_2AsH has been demonstrated from methanogenic bacteria, particularly

Methanobacterium strain Mo.H (Mo.H). Arsenate salts were the substrate and $MeCoB_{12}$ was a requirement for Me_2AsH production. Hydrogen, adenosine triphosphate (partially) and cell extract were also needed. Boiling prevented synthesis of Me_2AsH. This suggests the methylation works by a non-enzymatic transfer from Me-$CoCoB_{12}$ to arsenic. The original mechanism of Challenger was supported by this work. In addition with C.humicola, cell extracts and arsenate, $Me_2AsO(OH)$, $MeAs(O)(OH)_2$ and Me_3AsO are found i.e. the intermediates in Challenger's mechanism.[215-219] Arsenite, methylarsonate ($MeAsO_3^{2-}$) and dimethylarsinate ($Me_2AsO_2^-$) may also be methylated to Me_3As by C. humicola.

Such methylation under environmental conditions sheds light on to the organic arsenic derivatives found in seawater, marine biota and other waters. No free arsine derivatives seem to have been found in sea water to date; there is no evidence yet of biological volatilazation from surface waters but if there is it is likely that oxidation then takes place.

In a similar fashion to the mercury case, mixed bacterial cultures can biomethylate $Me_2AsO(OH)$ to arsenate.[220] This suggests a balanced arsenic cycle in the marine environment (Figure 1).

A number of organoarsenic compounds have been detected in oil shale retort and process waters, including $MeAsO(OH)_2$ and $PhAsO(OH)_2$. These were ascribed to prehistoric formation and fixing in the shale or they could have been formed chemically during the industrial process.[205]

Arsenic methylation from freshwater and other organisms has also been demonstrated recently with arsine or arsonic acid derivatures being identified.[99,221-222]

A number of other metals have been postulated to undergo environmental methylation. Some of the experimental evidence has been somewhat remote from environmental conditions (e.g. non-aqueous solvents). Here only those cases where reasonable environmental evidence exist are mentioned. Aqueous or sediment incubations leading to observable stable methyl metal species have been made for thallium (I), producing Me_2Tl^+ species,[223] selenium and tellurium species to the dimethyl compound.[224] The role of the anti-knock additive $MeCpMn(CO)_3$ in the environment has also been discussed.[225] Methyl antimony species have recently been detected in the environment.[226] There is some early laboratory evidence for the methylation of this element. Detection was by sodium borohydride reduction to methyl stibine derivatives followed by AA analysis. From natural waters methylstibonic acid ($MeSbO(OH)_2$) and dimethyl stibinic acid (Me_2SbOOH) were detected between the 1 and 13 ng dm^{-3} level. The origin of the methyl species was ascribed to biological methylation by algae[226].

Let it not be inferred that the environmental impact of organometallic compounds in general, and biomethylation in particular, has not been of historic significance. We should not forget the case of Napoleon I, Emperor of France. It has recently been suggested that Napoleon was probably poisoned by gaseous Me_3As emission from arsenic biomethylation of the wallpaper in his bedroom on St. Helena[227]. The green wallpaper has been shown to contain arsenic and the damp climate of St. Helena is suitable for the growth of moulds capable of biomethylation of arsenic to Me_3As. However this interesting hypothesis has also been disputed[228].

Acknowledgements

The author gratefully acknowledges funding over the past few years from the Natural Environment Research Council and the Science and Engineering Research Council. The receipt of fieldwork and travel funding from the Royal Society, the British Council and Leicester Polytechnic is also acknowledged with thanks.

REFERENCES

1. P.J. Craig, Environmental Aspects of Organometallic Chemistry, in Comprehensive Organometallic Chemistry, ed. E.W. Abel, F.G.A. Stone and G. Wilkinson, vol. 2, Pergamon Press, 1982, p. 979-1020.
2. The Biogeochemistry of Lead in the Environment (vols 1 and 2), ed. J.O. Nriagu, Elsevier North Holland,1978.
3. The Biogeochemistry of Mercury in the Environment, ed. J.O. Nriagu, Elsevier North Holland,1978.
4. Organotin Compounds: New Chemistry and Applications, ed. J.J. Zuckerman, A.C.S. Advances in Chemistry Series,157, A.C.S. Washington, D.C.,1976.
5. P.J. Craig, in Handbook of Environ. Chem., vol.1, part A, ed. O. Hutzinger, Springer Verlag, Berlin and Heidelberg, 1980, p.169.
6. Metal Pollution in the Aquatic Environment, ed. U. Forstner and G.T.W. Wittmann, Springer Verlag, Berlin, Heidelberg, New York, 1979.
7. The Enzymes, ed. P.D. Boyer, vol. IX Group Transfer, part B, 3rd edn., Acad. Press, New York, 1973.
8. Organometals and Organometalloids - Occurrence and Fate in the Environment, ed. F.E. Brinckman and J.M. Bellama, A.C.S. Symposium Series No. 82, A.C.S. Washington, D.C., 1978.
9. Trace Elements in the Environment, ed. E.L. Kothny, A.C.S. Adv.Chem.Ser. No. 123, A.C.S. Washington, D.C., 1973.
10. An Assessment of Mercury in the Environment, National Academy of Science, Washington, D.C., 1977.
11. Environmental Health Criteria 1, Mercury, World Health Organization, Geneva, 1976.
12. Environmental Health Criteria 3, Lead, World Health Organization, Geneva, 1977.
13. Environmental Health Criteria 15, Tin, World Health Organization, Geneva, 1980.
14. British Pharmacopoeia (Vols. 1 and II), HMSO, London, 1980.
15. Martindale, The Extra Pharmacopoeia, The Pharmaceutical Press, 27th edn., London, 1979.
16. British Pharmacopoeia (Veterinary), HMSO, London, 1977.
17. Mercury and the Environment - Studies on Mercury Use and Emission, Biological Impact and Control, Organization for Economic Cooperation and Development, Paris, 1974.
18. Mercury Contamination in Man and his Environment, International Atomic Energy Agency, Vienna, 1972.
19. R.M. Harrison and D.P. Laxen, Lead Pollution, Causes and Control, Chapman and Hall, 1981.
20. S.J. Blunden, L.A. Hobbs and P.J. Smith,'Environmental Chemistry', ed. H.J.M.Bowen (Specialist Periodical Reports). The Royal Society of Chemistry, London, Vol. 3 (in press).
21. S.J. Blunden and A.H. Chapman, Environ. Tech. Letters, 1982, 3 267
22. Organotin Compounds (Vols 1-3), ed. A.W. Sawyer, M. Dekker, New York, 1971-72.
23. R.C. Poller, The Chemistry of Organotin Compounds, Logos Press, London, 1970.

24. P.J. Craig, Environ. Tech. Letters, 1980, 1, 225.

25. R.H. Fish, E.C. Kimmel, J.E. Casida, in Ref. 4, p.197.

26. R.H. Fish, J.E. Casida and E.C. Kimmel, in Ref. 8, p.82.

27. E.C. Kimmel, J.E. Casida and R.H. Fish, J. Agric. Food. Chem., 1980, 28, 117.

28. E.H. Blair, Environ. Qual. Saf. Suppl., 1975, 3, 406.

29. C.J. Soderquist and D.G. Crosby, J. Agric. Food Chem., 1980, 28, 111.

30. C.J. Evans, Tin and Its Uses, No. 100, 1974, 3.

31. D. Barug and J.W. Vonk, Pestic. Sci., 1980, 11, 77.

32. W.R. Blair, J.A. Jackson, G.J. Olsen, F.E. Brinckman and W.P. Iverson, in Abstracts Intern.Conf. Heavy Metals Environ., Amsterdam, Sept. 1981, p. 235.

33. B. Sugavanam, Tin and Its Uses, No. 126, 1980, 4.

34. P.J. Smith and L. Smith, Chem. Br., 1975, 11, 208.

35. J.J. Zuckerman, R.P. Reisdorf, H.V. Ellis III and R.R. Wilkinson in Ref. 8, p.397.

36. P.J. Smith, Toxicological Data on Organotin Compounds, Intern. Tin Res. Inst. Pub. No. 538.

37. R.S. Braman and M.A. Tompkins, Anal. Chem., 1979, 51, 12.

38. V.F. Hodge, S.L. Seidel and E.D. Goldberg, Anal. Chem., 1979, 51, 1256.

39. Y.K. Chau, P.T.S. Wong and G.A. Bengert, Anal. Chem., 1982, 54, 246.

40. J.A. Jackson, W.R. Blair, F.E. Brinckman and W.P. Iverson, Environ. Sci. Technol., 1982. 16, 110.

41. H.J.M. Bowen, Trace Elements in Biochemistry, Acad. Press N.Y., 1966.

42. J.J. Zuckerman, R.P. Reisdorf, H.V. Ellis III and R.R. Wilkinson, in Ref. 8, p.411.

43. J.J. Zuckerman, R.P. Reisdorf, H.V. Ellis III and R.R. Wilkinson, in Ref. 8, p.388.

44. P.J. Craig and S. Rapsomanikis, J. Chem. Soc., Chem. Commun., 1982, 114.

45. Y.T. Fanchiang and J.M. Wood, J. Am. Chem. Soc., 1981, 103, 5100.

46. Y.K. Chau, P.T.S. Wong, O. Kramar and G.A. Bengert, Abstracts Intern. Conf. Heavy Metals Environ., Amsterdam, Sept. 1981.

47. P.J. Craig, S. Rapsomanikis and P.A. Moreton, Paper submitted for publication, 1982.

48. C. Huey, F.E. Brinckman, S. Grim and W.P. Iverson, Proc. Int. Conf. Trans. Persist. Chem. Aquat. Ecosystem, N.R.C., Ottawa, Canada, 1974, p.II-73.

49. F.E. Brinkman, J. Organometallic Chem. Library, 1981, 12, 343

50. J.S. Thayer and F.E. Brinckman, Adv. Organometallic Chem., 1981, 20, 314

51. L.E. Hallas, J.C. Means and J.J. Cooney, Science, 1982, 215, 1505.

52. H.E. Guard, A.B. Cobet and W.M. Coleman III, Science, 1981, 213, 770.

53. I.M. Robinson, in Ref. 2, p.99.

54. J.W. Robinson, E.L. Kiesel and J.A.L. Rhodes, Environ. Sci. Health, 1979, A14, 65.

55. R.M. Harrison, R. Perry and D.H. Slater, Atmos. Environ.,1974, 8, 1187.

56. B. Radziuk, Y. Thomassen, J.C. Van Loon and Y.K. Chau, Anal. Chim. Acta, 1979, 105, 255.

57. R.M. Harrison and R. Perry, Atmos. Environ., 1977, 11, 847.

58. R.M. Harrison, J. Environ. Sci. Health, 1976, A11, 419.

59. R.M. Harrison and D.P.H. Laxen, Atmos. Environ., 1977, 11, 201.

60. M.D. Baker, P.T.S. Wong, Y.K. Chau, C.I. Mayfield and W.E. Innes, Abstracts Intern. Conf. Heavy Metals Environ., Amsterdam, Sept. 1981, p.645.

61. A. Laveskog, in Proc. 2nd Intern. Clean Air Congress, ed. H.M. Englund and W.T. Beery, Washington, D.C. 1971, Acad. Press, N.Y., 1971, p.549.

62. L.J. Snyder, Anal. Chim. Acta, 1967, 39, 591.

63. L.J. Purdue, R.E. Enrione, R.J. Thompson and B.A. Bonfield, Anal. Chem., 1973, 45, 527.

64. S. Hancock and A. Slater, Analyst (London), 1975, 100, 422.

65. T. Nielsen, H. Egsgaard, E. Larsen and G. Scholl, Anal. Chim. Acta, 1981, 124, 1.

66. W.R.A. De Jonghe, D. Chakraborti and F.C. Adams, Environ. Sci. Technol., 1981, 15, 1217.

67. J.W. Robinson and D.K. Wolcott, Environ. Letters, 1974, 6, 321.

68. J.W. Robinson, L. Rhodes and D.K. Wolcott, Anal. Chim. Acta, 1975, 78, 78.

69. B.L. Gulson, K.G. Tiller, K.J. Mizon and R.H. Merry, Environ. Sci. Technol., 1981, 15, 691.

70. Lead and Health (The Lawther Report DHSS), HMSO, London 1980.

71. D. Turner, Chem. Br., 1980, 16, 312.

72. A.W.P. Jarvie, R.N. Markall and H.R. Potter, Environ. Res., 1981, 25, 241.

73. Y.K. Chau and P.T.S. Wong, in Ref. 8, p.39.

74. F. Huber, U. Schmidt and H. Kirchmann, in Ref. 8, p.65.

75. J.R. Grove, Intern. Experts Disc. Lead-Occurrence, Fate and Pollution in the Marine Environ., Rovinj, Yugoslavia, Oct. 18-22, 1977.

76. G.F. Harrison, Intern. Experts Disc. Lead-Occurrence, Fate and Pollution in the Marine Environ., Rovinj, Yugoslavia, Oct. 18-22, 1977.

77. P.J. Craig, in Ref. 1, p.1009 and references therein.

78. Y.K. Chau, P.T.S. Wong, G.A. Bengert and O. Kramar, Anal.Chem., 1979, 51, 186.

79. B.G. Maddock and D. Taylor, Intern. Experts Disc. on Lead-Occurrence, Fate and Pollution in the Marine Environ., Rovinj, Yugoslavia, Oct.18-22, 1977.

80. P.T.S. Wong, Y.K. Chau, O. Kramar and G.A. Bengert, Water Res., 1981, 15, 621.

81. P. Grandjean and T. Nielsen, Residue Rev., 1979, 72, 97.

82. G.R. Sirota and J.F. Uthe, Anal. Chem., 1977, 49, 823.

83. Y.K. Chau, P.T.S. Wong, O. Kramar, G.A. Bengert, R.B. Cruz, J.O. Kinrade, J. Lye and J.C. Van Loon, Bull.Environ.Contam.Toxicol., 1980, 24, 265.

84. B.A. Silverberg, P.T.S. Wong and Y.K. Chau, Arch. Environ. Contam. Toxicol., 1977, 5, 305.

85. D.C. Reamer, W.H. Zoller and T.C. O'Haver, Anal. Chem., 1978, 50, 1448.

86. R.M. Harrison and D.P.H. Laxen, Nature, 1978, 275, 738.

87. J.W. Robinson, E.L. Kiesel, J.P. Goodbread, R. Bliss and R. Marshall, Anal. Chim. Acta, 1977, 92, 321.

88. S.A. Estes, P.C. Uden and R.M. Barnes, Anal.Chem., 1981, 53, 1336.

89. Y.K. Chau, P.T.S. Wong and P.D. Goulden, Anal.Chim.Acta, 1976, 85, 421.

90. J.D. Messman and T.C. Rains, Anal. Chem., 1981, 53, 1632.

91. M.D. Dupuis and H.H. Hill,Jr., Anal. Chem., 1979, 51, 292.

92. P.T.S. Wong, Y.K. Chau and P.L. Luxon, Nature, 1975, 253, 263.

93. A.W.P. Jarvie, R.N. Markall and H.R. Potter, Nature, 1975, 255, 217.

94. A.P. Whitmore, A Study of Lead Alkylation in Natural Systems. Ph.D. Thesis, University of Aston, 1981.

95. P.J. Craig, Environ. Tech. Letters, 1980, 1, 17.

96. F. Huber, U. Schmidt and H. Kirchmann in Ref. 8, p.65.

97. J.-P. Dumas, L. Pazdernik and S. Belloncik, Proc. 12th Canad. Symp. Water Pollution, Res. Canada, 1977, p.91.

98. J.A.J. Thompson, Abstracts Intern. Conf. Heavy Metals Environ., Amsterdam, Sept. 1981, p.653.

99. M.D. Baker, P.T.S. Wong, Y.K. Chau, C.I. Mayfield and W.E. Innes, Abstracts Intern. Conf. Heavy Metals Environ., Amsterdam, Sept.1981, p.645.

100. F. Huber, U. Schmidt and H. Kirchman, in Ref. 8, p.66.

101. H. Agaki, Y. Fujita and E. Takabatake, Chem. Letters, 1975, 1, 171.

102. J.A.J. Thompson and J.A. Crerar, Mar. Pollut. Bull., 1980, 11, 251.

103. K. Reisinger, M. Stoeppler and H.W. Nurnburg, Abstracts Intern. Conf. Heavy Metals Environ, Amsterdam, Sept. 1981, p.649.

104. K. Reisinger, M. Stoeppler and H.W. Nurnburg, Nature, 1981, 291, 228.

105. W.P. Ridley, L.J. Dizikes and J.M. Wood, Science, 1977, 197, 329.

106. G. Agnes, S. Bendle, H.A.O. Hill, F.R. Williams and R.J.P. Williams, J. Chem. Soc., Chem. Commun., 1971, 850.

107. R.T. Taylor and M.L. Hanna, J. Environ. Sci. Health, 1976, A11, 201.

108. J. Lewis, R.H. Prince and D.A. Stotter, J. Inorg. and Nucl. Chem., 1973, 35, 341.

109. J.B. Honeycutt,Jr. and J.M. Riddle, J. Am. Chem. Soc., 1960, 82, 3051.

110. J.B. Honeycutt,Jr. and J.M. Riddle, J. Am. Chem. Soc., 1961, 83, 369.

111. Ref. 17, p.23.

112. Ref. 18, p.35.
113. K. Beijer and A. Jernelov, in Ref. 3, p.203.
114. Ref. 17, p.39.
115. N. Fimreite, in Ref. 3, p.601.
116. J.W. Huckabee, J.W. Elwood and S.C. Hildebrand, in Ref. 3, p.277.
117. Y. Takizawa, in Ref. 3, p.359.
118. T. Suzuki, in Ref.3, p.399.
119. Y. Kimura and V.L. Miller, J. Agric. Food Chem., 1964, 12, 253.
120. K. Furukawa, T. Suzuki and K. Tonamura, Agric.Biol.Chem., 1969, 33, 128.
121. Th.M. Lexmond, F.A.M. de Haan and M.J. Frissel, Neth.J.Agric.Sci., 1976, 24, 79.
122. K. Tonamura and F. Kanzaki, Biochim.Biophys.Acta, 1969, 184, 227.
123. G. Billen, C. Joiris and R. Wollast, Water Res., 1974, 8, 219.
124. F. Ribeyre, A. Delarche and A. Boudou, Environ.Pollut. (Ser. B), 1980, 1, 259.
125. N. Imura, S-K. Pan, M. Shimitzu and T. Ukita, in New Methods Environ. Chem.Toxicol., Collected Papers Res. Conf. New Methods Ecol.Chem., ed. F. Coulston, Int.Acad.Print.Co.Ltd., Tokyo, Japan, 1973, p.211.
126. N. Imura, S-K. Pan, M. Shimitzu, T. Ukita and K. Tonamura, Ecotoxicol. Environ.Safety, 1977, 1, 255.
127. J.W. Huckabee, J.W. Elwood and S.G. Hildibrand, in Ref. 3, p.283.
128. F. Frimmel and H.A. Winkler, Vom Wasser, 1975, 45, 285.
129. Y.K. Chau and H. Saito, Int.J.Environ.Anal.Chem., 1973, 3, 133.
130. J. Stary, B. Havlik, J. Prasilova, K. Kratzer and J. Hanusova, Int.J. Environ.Anal.Chem., 1978, 5, 84.
131. Y. Takizawa, in Ref. 3, p.325.
132. H.L. Windom and D.R. Kendall, in Ref. 3, p.303.
133. L. Friberg, Report Expert Group, Nord.Hyg.Tidskr.Suppl. 4, U.S. Andersons Tryckeri, Stockholm 1971.
134. P.D. Bartlett, P.J. Craig and S.F. Morton, Sci.Total.Environ., 1978, 10, 245.
135. P.D. Bartlett and P.J. Craig, Water Res., 1981, 15, 37.
136. P.J. Craig and P.M. Moreton, unpublished results, 1982.
137. P.J. Craig and S.F. Morton, Nature, 1976, 261, 125.
138. F. Glockling, Anal. Proc., 1980, 417.
139. P.D. Bartlett, P.J. Craig and S.F. Morton, Nature, 1977, 267, 606.
140. J.F. Uthe, J. Soloman and B. Grift, J. Assoc. Official Anal. Chem., 1972, 55, 583.
141. T. Fagerstrom and A. Jernelov, Water Res., 1971, 5, 121.
142. P.J. Craig and P.D. Bartlett, Nature, 1978, 275, 635.
143. K. Beijer and A. Jernelov in Ref. 3, p.208.

144. J.M. Wood, Environ. Sci. Res., 1980, 17, 223.

145. G. Westoo, Acta Chem. Scand., 1966, 20, 2131.

146. S. Jensen and A. Jernelov, Nature, 1969, 223, 753.

147. J.M. Wood, F.S. Kennedy and C.G. Rosen, Nature, 1968, 220, 173.

148. J.W. Vonk and A.K. Sijpesteijn, J.Ant.van Leeuwenhoek, 1973, 39, 505.

149. B.H. Olsen and R.C. Cooper, Nature, 1974, 252, 682.

150. B.H. Olsen and R.C. Cooper, Water Res., 1976, 10, 113.

151 M.K. Hamdy, O.R. Noyes and S.R. Wheeler, in Biol.Implics. Metals Environ., E.R.D.A. Symp.Ser. No. 42, 1977, p.20.

152. M.K. Hamdy and O.R. Noyes, Appl.Microbiol, 1975, 30, 424.

153. K.L. Jewett, F.E. Brinckman and J.M. Bellama, in Marine Chem. in the Coastal Environ., ed. T.C. Church, A.C.S. Symp. Ser.No. 18, A.C.S., Washington, D.C.,1975, p.304.

154. W.J. Spangler, J.L. Spigarelli, J.M. Rose, R.S. Flippin and H.M. Miller, Appl. Microbiol., 1973, 25, 488.

155. W.J. Spangler, J.L. Spigarelli, J.M. Rose and H.M. Miller, Science, 1973, 180, 192.

156. M. Yamada and K. Tonamura, J. Ferment. Technol., 1972, 50, 159.

157. M. Yamada and K. Tonamura, J. Ferment. Technol., 1972, 50, 893.

158. M. Yamada and K. Tonamura, J. Ferment. Technol., 1972, 50, 901.

159. I.R. Rowland, M.J. Davies and P. Grasso, Arch. Environ. Health, 1977, 32, 24.

160. S.E. Lindberg and R.C. Harriss, J. Water Pollut. Control Fed., 1977, 49, 2479.

161. J. Gavis and J.F. Ferguson, Water Res., 1972, 6, 989.

162 D.C. Gillespie, J.Fish.Res.Board Can., 1972, 29, 1035.

163. R.J. Pentreath, L.J. Exp.Mar.Biol.Ecol., 1976, 24, 103.

164. J.J. Bisogni, Jr., in Ref. 3, p.211.

165. J.J. Bisogni, Jr. and A.W. Lawrence, J. Water Pollut. Control Fed., 1975, 47, 135.

166. B.H. Olsen and R.C. Cooper in Ref. 150, p.113.

167. K. Beijer and A. Jernelov, in Ref. 3, p.205.

168. P.J. Craig and S. Rapsommonikis, in Environ.Spec. and Monitoring Needs Trace Metal-Containing Substances from Energy-Related Processes, proc. Do.E., N.B.S. Spec. Publ. No. 618, p.54., Washington D.C., 1981.

169. I.R. Rowland, M.J. Davies and P. Grasso, Nature, 1978, 265, 718.

170. G. Topping, Nature, 1981, 290, 243.

171. R.D. Rogers, U.S. Environ. Proc. Agency, Ecol. Res. Ser. Dept., 1977, EPA-600/3-77-007.

172. R.D. Rogers, J. Environ. Qual., 1977, 6, 463.

173. G. Lofroth, Ecol.Res.Bull. No.4, Swed.Nat.Sci.Res. Council, 1969.

174. S. Kitamura, Jumamoto Igk., 1963, 37, 494.

175. R.D. Rogers, Int.Conf.Heavy Metals Environ., Toronto, Oct. 1975, p.C218.

176. D.L. Johnson and R.S. Braman, Environ.Sci.Technol., 1974, 8, 1003.

177. R.D. Rogers, J. Environ. Qual., 1976, 5, 454.

178. D.G. Langley, J. Water Pollut. Control Fed., 1973, 45, 44.

179. H. Windom, W. Gardner, J. Stephens and F. Taylor, Estuarine Coastal Mar. Sci., 1976, 4, 579.

180. J.E. Blum and R. Bartha, Bull. Environ. Chem. Toxicol., 1980, 25, 404.

181. The Chemistry of Mercury, ed. N. McAuliffe, MacMillan, London, 1977.

182. Arsenical Pesticides, ed. E.A. Woolson, A.C.S. Symposium Series No. 7, A.C.S. Washington, D.C., 1975.

183. S.A. Peoples, in Ref. 182, p.1.

184. W.R. Penrose, C.R.C. Critical Revs. in Environ. Control, 1974, 465.

185. J.S. Edmonds and K.A. Francesconi, Mar. Poll. Bull, 1981, 12, 92.

186. M.O. Andreae, Deep Sea Res., 1978, 25, 391.

187. D.L. Johnson, Nature, 1972, 240, 44.

188. N.J. Blake and D.L. Johnson, Deep Sea Res., 1976, 23, 773.

189. D.L. Johnson and M.E.Q. Pilson, Environ. Letters, 1975, 8, 157.

190. A.W. Turner and J.W. Legge, Aust.J.Biol.Sci., 1954, 7, 452; 496; 504.

191. E.A. Woolson in Ref. 182, p.97.

192. S. Kurosawa, K. Yasuda, M. Taguchi, S. Yamazaki, S. Toda, M. Morita, T. Uehiro and K. Fuwa, 1980, Agric.Biol. Chem., 1980, 44, 1993.

193. R.V. Cooney, R.O. Mumma and A.A. Benson, Proc.Nat.Acad.Sci U.S.A., 1978, 75, 4262.

194. A.A. Benson, R.V. Cooney and J.M. Herrera-Lasso, J. Plant Nutrition, 1981, 3, 285.

195. T.J. Smith, E.A. Crecelius and J.C. Reading, Environ. Health Perspect., 1977, 19, 89.

196. E.A. Crecelius, Environ. Health Perspect., 1977, 19, 149.

197. A.E. Hiltbold, in Ref. 182, p.53.

198. E.A. Woolson and U.C. Kearney, Environ.Sci.Technol., 1973, 7, 47.

199. L.R. Johnson and A.E. Hiltbold, Soil Sci. Soc. Amer. Proc., 1969, 33, 279.

200. A.E. Hiltbold, B.F. Hajek and G.A. Buchanan, Weed Sci., 1974, 22, 272.

201. P.J. Ehman, Proc. S. Weed Conf., 1965, 18, 685.

202. C.C. Calvert, in Ref. 182, p.70.

203. R.S. Braman and C.C. Foreback, Science, 1973, 182, 1247.

204. M.O. Andreae, Anal. Chem., 1977, 49, 820.

205. R.H. Fish, F.E. Brinckman and K.L. Jewett, Environ.Sci.Technol., 1982, 16, 174.

206. R.S. Braman, D.L. Johnson, C.C. Foreback, J.M. Ammons and J.L. Bricker, Anal.Chem., 1977, 49, 621.

207. M.H. Arbab-Zovar and A.G. Howard, Analyst (London), 1980, 105, 744.

208. A.A. Grabinski, Anal. Chem., 1981, 53, 966.

209. Am.Pub.Health Assoc., Std.Methods Exam.Water Waste water, 13th edn., Washington, D.C., 1971.

210. G.M. George, L.H. Frahm and J.P. McDonnell, J.Assoc.Off.Anal.Chem., 1973, 56, 793.

211. E.B. Sandell, Colorimetric Determination of Traces of Metals, 3rd edn., Interscience, New York, 1959.

212. F. Challenger, Aspects of the Organic Chemistry of Sulphur, Butterworths, London, 1959.

213. F. Challenger, Quart. Rev. Chem. Soc., 1955, 9, 255.

214. D.P. Cox and M. Alexander, Applied Microbiol, 1973, 25, 408.

215. B.C. McBride and R.S. Wolfe, Biochemistry, 1971, 10, 4312.

216. B.C. McBride and T.L. Edwards, Biol.Implic.Metals. Environ., E.R.D.A. Symp. Ser. 42, Proc. Ann. Hanford Life Sci. 15th Symp.,1977, 1.

217. B.C. McBride, H. Merilees, W.R. Cullen and W. Pickett, in Ref.8, p.94.

218. W.R. Cullen, B.C. McBride and W. Pickett, Can.J.Microbiol., 1979, 25, 1201.

219. W.R. Cullen, B.C. McBride and M. Reimer, Bull.Environ.Contam.Toxicol., 1979, 21, 157.

220. J.G. Sanders, Chemosphere, 1979, 3, 135.

221. W.R. Cullen, C.L. Froese, A. Lui, B.C. McBride, D.J. Patmore and M. Reimer, J. Organometallic Chemistry, 1977, 139, 61.

222. P.T.S. Wong, Y.K. Chau, L. Luxon and G.A. Bengert, Proc. 11th Ann.Conf. Trace Subst.Environ.Health, Columbia, Mo., U.S.A., June 1977.

223. F. Huber, U. Schmidt and H. Kirchmann, in Ref.8, p.73.

224. Y.K. Chau, P.T.S. Wong, B.A. Silverberg, P.L. Luxon and G.A. Bengert, Science, 1976, 192, 1130.

225. M.D. Dupuis and H.H. Hill Jr., Anal.Chem., 1979, 51, 292.

226. M.O. Andreae, J.-F. Asmode, P. Foster and L. Van'tdack, Anal. Chem., 1981, 53, 1766.

227. D.E.H. Jones and K.W.D. Ledingham, Nature, 1982, 299, 626.

228. P.K. Lewin, R.G. Hancock and P. Voynovitch, Nature, 1982, 299, 627.